Medical Cannabis
The Definitive Guide

David A. Dawson, PhD

Copyright © Dr. David A. Dawson, PhD

All rights reserved. No part of this book other than the Table of Contents may be reproduced in any form or by any electronic or mechanical means, including information storage and retrieval systems, without permission in writing from the author and publisher, except by reviewers, who may quote brief passages in a review.

ISBN: 979-8-218-49131-4

DEDICATION

This book is dedicated to all the people of the world who search for and disseminate truth despite the institutionalized barriers to doing so imposed by society.

Table of Contents

About the Author 5

Introduction 8

The Convoluted Nature of Cannabinoid Nomenclature 10

The History of Medicinal Botanic Cannabinoid Use 13

The Institutionalized Stigmatization of Botanic Cannabinoid Medicines 41

Cannabis Chemovars and a Holistic Perspective of Why Cannabis Works as Medicine 62

The Relationship of the Endocannabinoid System to Autoimmune Disorders and Addiction 68

The Endocannabinoid System and its Relationship to Dosing with Cannabis 77

Using Cannabis Strain Fingerprints to Target Specific Medical Conditions 83

Methods of Ingestion of Medicinal Cannabis 116

Diagnosis and Treatment of Endocannabinoid Deficiency Disorders: A Study Proposal 130

Specific Medical Conditions and Their Treatment with Medicinal Cannabis 133

About the Author

The history of scientific discovery is abundant with what are sometimes considered to be serendipitous circumstances, and mine is a story of serendipity. It's a story of how the protocols of the American Healthcare delivery system sucked me in, chewed me up, and spit me into the field of biomolecular psychology. This story illustrates the protocols of the American Healthcare Delivery system. I don't have any firsthand knowledge of the healthcare systems of other countries, but I understand the subtleties of the American healthcare delivery system rather well because this system has been intimately engaging me for more than five decades.

When I was four years old, I was diagnosed with a form of muscular dystrophy known as Charcot-Marie-Tooth disease. Charcot-Marie-Tooth is a type of peripheral neuropathy that affects the transmission of information between the central nervous system and the rest of the body. Like other forms of muscular dystrophy, Charcot-Marie-Tooth is an extremely painful neuropathic condition. Typically, pain symptoms manifest between the ages of 5 and 25, and the condition is slowly progressive. In my case, neuropathic pain began around the age of eighteen. Opioids have for millennia been regarded as the most effective synthetic molecules for treating neuropathic pain, and following the protocol of the American Healthcare Delivery System, my doctors prescribed codeine.

Codeine worked great for about two years until I developed a tolerance. And following established medical protocol, my doctors exclaimed, "That's okay! We have plenty of opiates to choose from. The protocol is to change the opiate and increase the dose." My father was an anesthesiologist, so I was conditioned as a child to adhere to medical protocols, and as a result, throughout my adult life, I was prescribed and ingested increasing amounts of Oxycontin, Hydrocodone, Oxycodone, Vicodin, Codeine, Opana, and Fentanyl, developing a tolerance to all, and finally ending up being prescribed 500 mg of Morphine three times a day. This dose caused me to suffer a stroke, which resulted in the loss of the use of my left arm. That was also the last day I walked, and I was placed in a Hospice facility where my doctors hovered around my deathbed, saying, "We've done everything we could for him. We followed our protocol of treating his pain by addicting him to opiates for his entire adult life."

At this point, my doctors explained to me that I had used up every oral opiate invented, and the protocol was now to put me into hospice because I had less than six months left to live. They assured me that I would go out comfortably, as the protocol now was to manage my pain by continuously increasing the intravenous morphine over those months, and eventually, I would stop breathing. This was the protocol they were institutionally mandated to follow by the hospital they were affiliated with and the insurance company with which I was affiliated.

At the time, I wasn't particularly interested in dying, so I started researching alternatives. One of the alternatives I researched was medicinal cannabis. I hadn't consumed any cannabis since college, and unlike Bill Clinton, in college, I inhaled deeply and frequently. At that time, I became aware that there was, in addition to the usual effects of ingesting botanic cannabinoids, a reduction in my pain.

I informed my doctors that I understood the paradigm under which they were operating and that the institution that makes their medical decisions for them provided them with no other options but that I had been following the pharmaceutical paradigm for almost five decades, and this is where it had gotten me. So, against my doctors' orders, I chose to try the nutraceutical approach, and the first nutraceutical I chose to try was medicinal cannabis. Their response was, "Oh, no, no, no! That has no medical applications whatsoever." To which I responded, "Well, I've been following the pharmaceutical protocol my entire life, and the protocol now is to slowly euthanize me."

I was able to obtain a supply of medicinal-grade cannabis. The medicine worked. It treated the neuropathic pain from the muscular dystrophy, in addition to the four-decade opiate addiction inflicted on me by the American Healthcare Delivery System. When I realized it worked, I felt that I had to figure out how it worked. I have a strong background in molecular biology, so I chose to enter a PhD program focusing on biomolecular psychology. The simplest way of conceptualizing biomolecular psychology is that it analyzes concepts holistically from a biomolecular perspective. Albert Einstein is credited with the idea that if you can't explain your concept simply, you simply don't understand your concept, and this book is designed to explore the principles of biomolecular psychology and how these principles apply to the science of medicine as it pertains to the medicinal molecules altruistically provided to humanity by the species *Cannabis sativa.*

Introduction

This book has an interesting backstory that illustrates the institutionalized stigmatization of scientists who pursue truth and disseminate research-based information pertaining to the medicinal properties of botanic cannabinoids and terpenes produced by the species *Cannabis Sativa*. Medical Cannabis: The Definitive Guide was originally published in 2016 and was designed to provide information to medical cannabis patients about the endocannabinoid system and how to utilize the medicinal properties of botanic cannabinoids and terpenes effectively.

The book was well received, physicians began using it to train their medical cannabis patients, and I was fortunate to be invited to speak at a few medical conventions. Enough physicians and clinicians attending these conventions bought this book to allow me to pay off a thirty-six hundred Bank of America credit card bill. Upon entering the bank and attempting to pay off that bill, a bank officer and a big, burly guard confronted me and escorted me into a tiny room. The officer politely informed me that because I had written a book detailing the medicinal properties of the botanic cannabinoids and terpenes the cannabis plant produces, their processor had determined that I was involved in the marijuana industry. As such, the financial institution could not accept payments on my credit card because doing so would subject them to money laundering. I was also, at that time, informed that this decision could not be appealed. Every month for the next five years, Bank of America reported my failure to make any payment on the credit card bill to the three credit bureaus. After five years of Bank of America destroying my credit rating for not making the payments on my credit card two days before Christmas in 2020, this institution formally sued me for the money that they refused to accept.

I was then introduced to the American Judicial System as an institution. I spoke with multiple attorneys who expressed their unwillingness to oppose lawyers representing Bank of America. I was informed that this institution had the greatest attorneys in the world, who demonstrated their prowess by convincing the United States to ignore the criminal activities this institution engaged in that caused the 1988 financial crisis. Every lawyer I consulted confessed they could not accomplish this feat for a client and did not feel comfortable opposing lawyers who had. I was forced to represent myself against these attorneys in Florida Case # 19-0001511-SC. The ruling of the court was that the law against writing books related to this topic had been adjusted with the 2018 ratification of the Farm Act, and it was my responsibility to inform Bank of America at that time that writing a book detailing the medicinal properties of botanic cannabinoids was no longer illegal in the United States of America and the Court ruled in favor of the plaintiff. My experience with this financial institution and the American Judicial System revealed the institutionalized stigmatization of researchers and authors who write books that delineate the medicinal properties of botanic cannabinoids and terpenes.

This book is an updated version of the 2016 edition of Medical Cannabis: The Definitive Guide. The 2016 edition was written before I attained my Doctorate of Psychology. This edition enhances the first in that it includes the information attained in my doctoral research focusing on biomolecular psychology, psychopharmacology, and psychoneuroimmunology. This is an adapted version of the previous edition and provides more extensive research-based information pertaining to the biomolecular mechanisms that result in the efficacious treatment of multiple ailments through the judicious administration of botanic cannabinoids and terpenes.

Chapter 1

The Convoluted Nature of Cannabinoid Nomenclature

Nomenclature is a system of applying names for terms used in a particular field of science. As in any science, the terms used in biomolecular psychology pertaining to cannabinology must be defined. Fortunately, the terms used in biomolecular psychology to classify cannabinoids are simple and only become convoluted when examined psychosocially.

The field of biomolecular psychology describes two types of cannabinoids: natural and synthetic. Synthetic cannabinoids are artificial cannabinoids that are chemically produced by scientists in a laboratory and designed to mimic the medicinal and intoxicative effects of botanic cannabinoids. Multiple examples of synthetic cannabinoids have FDA approval, and because they are FDA-approved, these synthetic molecules are considered efficacious for treating various ailments by the American Healthcare Delivery System. The paradigm under which the American Healthcare Delivery System currently operates is that only synthetic cannabinoids have efficacy because these molecules have attained the glorifying label of FDA-approved.

Synthetic Cannabinoids

Employers that screen employees for cannabis use do not test for the presence of synthetic cannabinoids. This has resulted in increased recreational use of synthetic cannabinoids in recent years. As this book was approaching publication, a group of scientists published a paper in the European Journal of Medical Research delineating the health risks and toxicity of synthetically produced cannabinoids (Alzu'bi et al., 2024). A multitude of reports have linked synthetic cannabinoid consumption to the incidence of various adverse health effects, turning their widespread use into a major public health concern. These compounds are chemically designed to mimic the intoxicating effects of $\Delta 9$-tetrahydrocannabinol (THC), the main intoxicating ingredient of *Cannabis sativa* (Castaneto et al., 2014; Fattore & Fratta, 2011), and most synthetic cannabinoids are significantly more intoxicating than the THC produced naturally by the cannabis plant (Gioé-Gallo et al., 2023; Murry et al., 2016). In addition to binding to the same receptors as naturally produced cannabinoids, it has been demonstrated that synthetic cannabinoids also interact with non-cannabinoid targets, resulting in distinct pharmacologic effects as well as a diverse toxicity profile (De Petrocellis & Di Marzo, 2010; Hess et al., 2016).

Numerous studies and reports link synthetic cannabinoid ingestion with the emergence of a wide range of serious adverse health effects. These effects are not limited to the central nervous system but are also exhibited in other body targets, including cardiovascular, renal, respiratory, digestive, and immune systems. The most commonly reported toxic effects linked to synthetic cannabinoid ingestion include agitation, anxiety, drowsiness, nausea, vomiting, depressed breathing, tachycardia, hypertension, muscle twitches, as well as more dangerous effects, such as psychosis, cognitive impairment, stroke, seizures, cardiac complications, acute renal failure, and acute hepatic injury (Alipour et al., 2019; Alves et al., 2020; Castaneto et al., 2014). Moreover, reports of overdose deaths following the ingestion of synthetic cannabinoids have increased remarkably in recent years (Adamowicz, 2016; Armstrong et al., 2019; Kasper et al., 2019; Shanks, 2016). The mechanism by which synthetic cannabinoid ingestion results in toxic effects is discussed in later chapters, particularly Chapter 3.

Natural Cannabinoids

Two types of natural cannabinoids exist. Phytocannabinoids (botanic cannabinoids) and endocannabinoids. Phytocannabinoids (Phyto means flower) are the cannabinoids produced by the cannabis plant. Endocannabinoids (Endo means from within) are the cannabinoids produced by the vertebrate body. Intriguingly, the endocannabinoids and phytocannabinoids are the same molecules. Or, if not technically the same molecules, they act on the body's receptors in the same way. This will be discussed in greater detail in later chapters.

The equivalency of endocannabinoids and phytocannabinoids provides a mechanism for explaining the medical efficacy of phytocannabinoid supplementation for endocannabinoid deficiency disorders. This equivalency is relevant to autoimmune disorders, as a systematic review conducted by Katchan et al. (2016) demonstrates that virtually every known autoimmune disorder is an endocannabinoid deficiency disorder. The theory of phytocannabinoid supplementation for endocannabinoid deficiency disorders predicts that the deficiency can be effectively treated through the supplementation of the equivalent botanic cannabinoid. For example, evidence indicates that deficiencies in anandamide can be effectively treated through the supplementation of Δ9-Tetrahydrocannabinol, deficiencies in the endocannabinoid 2-arachidonoylglycerol (2AG) can be effectively treated through the supplementation of cannabidiol, and a deficiency in Virodhamine may be effectively treated through the supplementation of the botanic cannabinoid tetrahydrocannabivarin.

This is the point where things become convoluted psychosocially because, in the United States, politicians make medical decisions and determine which aspects of truth scientists are permitted to pursue. In 1972, President Richard Nixon prohibited scientists residing in the United States from pursuing knowledge pertaining to any cannabinoid that was not produced synthetically. Ironically, through the National Institute of Health (NIH), the United States funded studies of botanic cannabinoids conducted in other countries (Devane et al., 1992), and because of the Nixon Mandate, these countries utilize phytocannabinoid supplementation to alleviate endocannabinoid deficiency disorders routinely and are arguably about half a century ahead of scientists in the United States in their understanding of this next frontier of medicine.

Chapter 2

The History of Medicinal Botanic Cannabinoid Use

Medicinal botanic cannabinoid use has a long and convoluted history. Crocq (2020) retraced the 12,500-year history of therapeutic botanic cannabinoid use from the earliest contact of humans with the cannabis plant at the end of the ice age to its subsequent global expansion, its medicinal uses, and the discovery of the endocannabinoid system in the 20th century. Small (2015) detailed that because humans encountered cannabis before they began recording their history, the evidence of their initial encounter is indirect, and there can be no firm answers to questions about the origins of human's initial associations with this plant. Still, it is possible to formulate logical suppositions with the aid of various sets of information.

Chinese archaeologists Jiang et al. (2013) and Zhang et al. (2017) have firmly concluded that the Yang Shao culture was established precisely where cannabis originated, just south of Siberia, and cannabis use extends deep into our prehistorical human past. An accumulation of palaeobotanical evidence in the form of pollen deposits (Herbig & Sirocko, 2013), archeological evidence (Jiang et al., 2013), and historical accounts (Bradshaw et al., 1981; Duvall, 2014; Li, 1973; Murphy et al., 2011) strongly suggests an evolving symbiotic relationship with the cannabis plant that remarkably synergistically enhanced the human civilization process. Research of the evolutionary forces that affected the development of domesticates by Rindos (1984) implied cannabis plants established a symbiotic, co-evolutionary relationship with protohumans, beginning with early humans merely protecting wild cannabis and progressing until its domestication improved the plant and the humans utilizing the resources it provided, which helped each increase in numbers. Rindos (1984) suggested that early humans and plants manipulated the other's environment to produce genetic changes in their DNA, favoring reproduction using the least amount of energy.

Rindos (1984) implied that this symbiotic relationship offered a substantial evolutionary advantage to both species. He purported that from a genetic perspective, plants are programmed by their DNA to make seeds and spread them. Similarly, humans are evolutionarily programmed to reproduce and increase in numbers. The symbiotic relationship benefited both species in this arrangement as it enabled both the plant and the animal to accomplish this primary task. Rindos (1984) theorized that once plant seeds are produced, a genetic algorithm will always find the most efficient dispersal method, be it animal fur/feces, water, wind, or by influencing human behavior. In exchange for providing its medicine (the plant's gift to humans), protohumans were only too happy to shelter it and provide it with nutrients, water, and protection from predators and the elements—a win-win. The plant also provided humans with many necessary survival products: a high-quality complete food containing essential fatty acids, oil for cooking and lamps, lubrication for skin and hair care products in addition to clothing, and naturally salt-resistant rope, which was extremely durable.

Using large-scale whole-gene resequencing, Ren et al. (2021) demonstrated that cannabis was first domesticated in early Neolithic times in East Asia and that all existing cannabis and hemp cultivars diverged from an ancestral gene pool currently represented by feral plants and landraces in China. Beaumont and Li (2022) depicted hemp farming as initiating and accelerating the first facets of human civilization. Crandall (2020) described cannabis as a 'first foundation' crop that gave protohumans an evolutionary advantage by producing fiber, food, fuel, and raw building material. As well, research on the evolution of *Cannabis sativa* and hemp by Ren et al. (2021) and Small (2015) suggested that human domestication of the cannabis plant played a pivotal role in the initiation and development of human civilization.

Civilization implies cooperation over hostile tribal conflicts. Over time, this plant yielded something new: a 'medication for the soul'—a plant that could satisfy a unique higher purpose, the ability to positively change human consciousness. This is the fourth drive in Maslow's (2017) hierarchy of human needs after food, shelter, and reproduction. The amalgamation of this research suggests that for early humans, discovering cannabis/hemp could arguably be considered analogous to hitting the evolutionary lotto.

Subsequent secondary domestication events in non-native geographical regions accelerated the civilization process by causing a radical shift in human behavior called the Neolithic Revolution. The Neolithic Revolution was a period of significant transition where humans transitioned from hunter-gatherer subsistence to primitive agricultural settlements. Lu and Clark (1995) suggested that the first agrarian settlements likely originated as hemp/cannabis-based farms, where humans began to settle down in one place with a single goal: finding a niche with an optimal environment to grow plants for food. An economic analysis by Putterman (2008) demonstrated that the Neolithic Revolution occurred in six different early cultures worldwide in 5000 years, but the first area in the world where it took place was in northern China, very close to where wild hemp originated. Human's first cultivation of hemp also started there. Using a synthetic approach in their investigations of pollen, charcoal, and seed remains, Wang et al. (2019) demonstrated that the earliest known farming culture to use hemp/cannabis was the Yang Shao culture, whose origins go back to 5000 B.C. This culture existed until 3000 B.C., and for those 2000 years, the economy of this culture was cannabis-driven, with excess cannabis being traded with other tribes. Jia et al. (2013) also revealed that China is the oldest culture in the world, and the export of hemp seeds and products progressed west and initiated the development of new Neolithic Revolutions in India and then the Middle East and Europe. This suggests that cannabis hemp played a substantial role in a critical aspect of civilization: the founding of human commerce.

Research by Kuddus et al. (2013) demonstrated that as soon as cannabis arrived in these regions and cultures, the development of cannabis-based religions occurred in these areas. Shamans from these cultures likely used cannabis as a sacred plant of peace and as something to worship. Underhill (1997) described the recovery in the Neolithic Yang Shao culture site of Banpo, a pottery bowl with hemp impressions. Furthermore, Jiang et al. (2013), Mukherjee et al. (2008), and Russo et al. (2008) described the discovery of remarkably well-preserved cannabis flowers, leaves, stems, and seed remains dated to about 2700 BP contained in two vessels in the Yanghai Tombs near Turpan in Western China, in a grave of a man that was identified as a shaman.

Another critical aspect of human civilization, the recording of its history, was also synergized by hemp. Duvall's (2014) research revealed that the earliest South Asian texts were written in Sanskrit and provided two plant names. Both translated as 'cannabis.' The first described a plant used to make cordage, textiles, and food; the second detailed plant medicines derived from cannabis. Archaeological research by Brand & Zaho (2017) revealed that approximately 10,000 years ago, humans began to write information on hemp paper as hemp replaced bulky clay tablets and expensive silk. Hemp paper became available to all people. It was exceptionally durable and easy to make in copious amounts, making the first books possible. This literally initiated humanity's first 'information age.'

Hemp, the plant, also initiated the civilization of medicine by providing a means of recording what happened in the past from which others could learn. Brand and Zaho (2017) revealed that the first books were made from hemp paper, that they were medical journals printed in China, and that the first information written in them was the suggested medicinal uses of cannabis. In addition to Brand and Zaho's (2017) research, Pisanti and Bifulco (2019) demonstrated that an abundance of written records of medicinal cannabis use is found in ancient Chinese texts. The emperor Shen Nung (2700 BCE) is considered the patron of all herbalists and apothecaries and provided the first written evidence of medicinal cannabis use in the most ancient Chinese Pharmacopoeia, the 'Shen Nung Pen Tsao Ching' written in the first century BCE. This document reported cannabis remedies traditionally used and orally handed down for over two thousand years from Emperor Shen Nung's reign.

Julien (1894) cited the handwritten reports on hemp paper of the famous surgeon Hua Tuo (110–207 CE), who performed surgical operations without causing pain to his patients upon the administration of oil made by a mixture of cannabis resin, Datura, and wine. A further demonstration of the dramatic influence of cannabis on the civilization of medicine is provided in an analysis by Jiang et al. (2013) of the desiccated and charred plant remains at the site of Yuergou, Xinjiang, China. This analysis demonstrated that hemp was one of the first known plant species to be grown purposefully and the only one cultivated dioeciously, meaning it has separate male and female plants. This is intriguing because it provides insight into what Neolithic people were thinking about cannabis and how they used it. There is only one use for the female cannabis plant: medication.

Upon its introduction into India, about 1000 years BCE, the therapeutic use of cannabis spread rapidly. Chopra and Chopra (1957) described the sacred virtues the Indian people attributed to this plant. The Indian people considered cannabis a source of happiness, capable of eliciting joy and feelings of freedom. The sacred bhanga, as it was named, was considered ideal for relieving anxiety, although it was not clearly separated from religious and ritual practices in Tibet. Cannabis was considered a sacred plant and was used in Tantric Buddhism to facilitate meditations in addition to its application in traditional Tibetan medicine. Knörzer (2000) described *Cannabis indica* as being used for its anesthetic, analgesic, antispastic, antiparasitic, and diuretic properties, as an expectorating agent, to treat convulsions, as an aphrodisiac, to stimulate hunger, and to provide relief from fatigue.

Cannabis was also used therapeutically by the ancient Egyptians, and medicinal cannabis was described in the Egyptian Pharmacopoeia since pharaonic times. Russo (2007) depicted Egyptians utilizing multiple modes of administration, orally, by rectum, vaginally, through fumigation, into the eyes, or topically. As well, Mahdizadeh et al. (2015) speculated that cannabis was brought to Arab countries very early by Arab travelers and sea traders arriving directly from India and detailing the Arabian use of cannabis for the treatment of headaches, degenerative bone, and joint diseases, ophthalmitis, general edema, gout, wounds, and uterine pain.

While cannabis was known in the Greek Latin Age, there is little evidence of it being used in medicine in the early times. Latins and Greeks primarily used it as fiber for rope and sailcloth. It is not until the first century CE that Greek medicine reflects the use of cannabis for its analgesic properties. Bonini et al. (2018) and Pisanti and Bifulco (2019) revealed that the writings of the Greek physician Pliny depicted both positive and negative properties of cannabis, to which he attributed the ability to extract worms and parasites from the ears, relax contractions of the joints, cure gout, and alleviate burns but also identified contraindications including impotence in men and headache.

Based on palaeobotanical evidence, Sorenson and Johannesse (2009) postulated that the most likely mode of diffusion of cannabis into Europe and the Mediterranean basin was through Scythians or proto-Scythian people moving from Central Asia through Russia about 3,500 years ago.

Remnants of cannabis found in Scythians' graves in Germany, Siberia, and Ukraine date back to 450 BCE. Emboden (1972) revealed that the Scythians mainly used hemp during funeral proceedings and in banquets through fumigations, relaxing saunas, and as fiber and speculated that the Scythians could have learned of therapeutic cannabis use from Assyrians and Thracians in the region of the eastern Balkan peninsula and Dacia around 2600 BP.

Crocq (2020) described a significant tradition of medicinal cannabis use developing in Eastern Europe in the 16th century, with hemp flower routinely being mixed with olive oil and applied as a bandage on wounds or combined with hempseed oil to treat rheumatism and jaundice. Bifulco et al. (2015) detailed treatments devised by Polish botanist Symon Sirenius throughout the 16th century and his use of cannabis resin as a medicament to treat burns and hemp roots boiled in water as a treatment to relieve joint pain. One of the more interesting uses of cannabis in Polish popular medicine during this time was its utilization in the elimination of so-called tooth worms that, in those times, were considered the origin of tooth decay. The treatment entailed boiling hemp seeds in a pot and the vapor being inhaled by the patient. It was believed that this hemp porridge would intoxicate the worms and cause them to fall out. The amalgamation of these studies implies that virtually every aspect of the human civilization of medicine was profoundly synergized by its association with cannabis hemp.

The modern history of medical cannabis began in the 19th century when Irish physician William Brooke O'Shaughnessy applied scientific methodology for the first time to study the cannabis plant's pharmacological and toxicological properties. O'Shaughnessy conducted his experiments in India. MacGillivray (2017) described O'Shaughnessy dispersing his knowledge of the innumerable pharmacological benefits of cannabis to the European medical community upon his return to Europe. O'Shaughnessy's (1839) essays are abundant in historical details of the medicinal uses of cannabis as reported by Sanskrit, Persian, and Arab authors and consisted of pharmaceutical research subdivided into preclinical safety and efficacy studies followed by clinical experimentations. He recognized that the cultivars of cannabis grown in India *(Cannabis indica)* were different from the European varieties *(Cannabis sativa),* which were primarily used for manufacturing fiber rather than for their pharmacological properties. He rigorously approached the investigations on the cannabis extracts by employing in vivo research utilizing different test species, including mice, dogs, cats, and rabbits (O'Shaughnessy, 1839). Upon establishing the safety of the *Cannabis sativa* cultivar, he prepared alcoholic tinctures of the strain that he administered to some of his patients who were afflicted with rheumatism, cholera, and tetanus and to an infant suffering from convulsions (O'Shaughnessy, 1843). These patients responded well to the therapy, so he determined that the cannabis extract had potent analgesic and myorelaxant properties and proposed it as a remedy for seizures, leading him to profess that the clinical studies he conducted demonstrated that in cannabis the medical profession had gained an anticonvulsive remedy of the greatest medicinal value. The studies O'Shaughnessy (1843) performed on the effects of cannabis on tetanus led him to purport that even if it could not cure the infection, it proved useful as a palliative therapy to relieve severe symptoms. Booth (2005) portrayed the knowledge imparted by O'Shaughnessy to Britain about the medicinal and psychotropic properties of cannabis proliferating and the utilization of these palliative properties spreading throughout the European medical community. Still, attempts by eminent chemists to identify and isolate the medicinal molecules from alcoholic cannabis tinctures provided by O'Shaughnessy proved unsuccessful.

Mechoulam and Gaoni (1967) recounted multiple attempts to identify and isolate the medicinal molecules from O'Shaughnessy's alcoholic cannabis tinctures. A methodology problem of how plant substances were isolated at the time was described by Drobnik and Drobnik (2016), providing insight into why the first botanic cannabinoid was not identified until the middle 1900s. In the 1800s, distillation was a well-established shortcut for isolating bioactive plant products, with menthol having been purified as early as 1771. However, cannabinoids are poorly volatile compounds with comparable vaporization temperatures, and their distillation requires high vacuum and low-resolution power. It is, therefore, predictable that the numerous eighteenth-century investigations on cannabis would have markedly missed the identification of its medicinal compounds.

The stigmatization of cannabis in Western medicine began in the late 19th and early 20th centuries. Crocq (2020) speculated that a significant obstacle to the usage of cannabis was that the active ingredient had not been isolated, so plant extracts could not be made uniform. Reynolds (1890) identified variations in the therapeutic agent of hemp grown during different seasons and places, an issue absent in medicines created synthetically.

In the late 1930s and early 1940s, the botanic cannabinoids cannabinol (CBN) and cannabidiol (CBD) were isolated by Adams (1940) from hemp oil, and CBD was isomerized into a mixture of two tetrahydrocannabinols with "marihuana-like" physiological activity in dogs and establishing their structure except for the location of one double bond. Two years later, Wollner et al. (1942) isolated Δ9-Tetrahydrocannabinol (Δ9-THC) from cannabis resin. Scientific advancements often depend on the development of instrumentation. In 1964, Gaoni and Mechoulam (1964) resolved the structure of Δ9-THC by uniting NMR spectroscopy with nuclear magnetic resonance imaging, thereby firmly establishing the position of the elusive double bond.

Slaughter (1988) recounted President Richard M. Nixon instituting a ban on the research of biologically produced cannabinoids in the United States on June 18, 1971, as part of his war on drugs. At the time, the only known biologically produced cannabinoids were the botanic cannabinoids produced biologically by the cannabis plant. In 1973, Nixon established the Drug Enforcement Administration (DEA), a paramilitary division of the Department of Justice, and tasked it with ensuring that no research of biologically produced cannabinoid molecules occurred within the confines of the United States of America.

Cannabinoid and cannabis research throughout the next two decades was portrayed by Dr. Raphael Mechoulam in a conversation as a rather esoteric field, only involving a small number of scientists in the United States and abroad. He depicted pursuits of truth in this area of inquiry as being constrained by the politicized agenda of the National Institute of Health (NIH) and the National Institute of Drug Abuse (NIDA), which had institutionalized programs ensuring subsidized studies of botanic cannabinoids were designed to demonstrate deleterious effects of botanic cannabinoid molecules while blocking inquiries into any potential health benefits.

In the late 1980s and 1990s, because pursuing knowledge related to these controversial molecules was prohibited in the United States, the National Institute of Health provided funding to William Devane and other researchers at Hebrew University in Jerusalem. Devane et al. (1992) discovered the endocannabinoid system using newly designed technology. They described this biological system as comprised of endogenous lipid-based retrograde neurotransmitters that bind to cannabinoid receptors and cannabinoid receptor proteins expressed throughout the vertebrate central and peripheral nervous systems. This group identified the first endogenous cannabinoid neurotransmitter molecule (endocannabinoid) and named it anandamide. Ananda is a Sanskrit word meaning bliss, referencing the euphoric feeling resulting from the Δ9-THC cannabinoid agonizing the CB1 receptors. In 1993, Mechoulam et al. (1995) discovered a second endocannabinoid molecule, 2-Arachidonoylglycerol (2AG).

Rather than discrediting cannabis, this research inadvertently facilitated a series of significant discoveries about the workings of the vertebrate brain (Maccarrone, 2022). These breakthroughs spawned a revolution in medical science and a profound understanding of health and healing. Iannotti et al. (2016) demonstrated that botanic cannabinoids Δ9-THC and CBD activate cannabinoid type 1 and type 2 receptors (CB1 and CB2), respectively. As well, Stasiłowicz et al. (2021) described botanic cannabinoid receptor agonism as being clinically efficacious for the suppression of nausea and emesis, as appetite stimulants, and as analgesics for neuropathic pain in adults with multiple sclerosis and advanced cancer.

Ethan B. Russo (2004) proposed the concept of clinical endocannabinoid deficiency disorders underlying the pathophysiology of migraine, fibromyalgia, irritable bowel syndrome, and other functional conditions alleviated by the intromission of botanic cannabinoids. Russo theorized that each neurotransmitter system could have pathological conditions caused by an endocannabinoid deficiency. For example, Alzheimer's dementia is attributable to loss of acetylcholine activity, Parkinsonism is attributable to dopamine deficiency, and depression is associated with reduced serotonin concentrations. This implies that endocannabinoid deficiencies, either congenital or acquired, explain the pathophysiology of these elusive conditions, particularly where the endocannabinoid receptors are especially dense. More research is required in this field, but a systematic review by Katchan et al. (2016) indicated that virtually every identified autoimmune disorder is likely the result of an endocannabinoid deficiency.

Dawson (2018) applied this idea to the potential development for treating obesity and type II diabetes, identifying endocannabinoid/ phytocannabinoid equivalents and the concept of phytocannabinoid supplementation for treating endocannabinoid deficiency disorders. He reviewed numerous studies by the pharmaceutical company Sanofi-Aventis demonstrating that central cannabinoid (CB1) receptors play a significant role in controlling food consumption and dependence. To develop suitable synthetic medicines against this target, Sanofi-Aventis screened compounds with potential activity against this receptor for inhibitory activity, resulting in the pharmaceutical company creating a synthetic cannabinoid called Rimonabant as a potent CB1 receptor antagonist. Preclinical animal trials demonstrated that Rimonabant reduced the consumption of fats and sugars, which are significant contributors to weight gain.

Manipulation of CB1 receptors with Rimonabant by Stanley et al. (2013) resulted in a significant reduction in body weight, waist circumference, triglyceride concentrations, an increase in HDL cholesterol and adiponectin concentrations, and a reduced number of subjects with type 2 diabetes. These preclinical findings were confirmed in a series of clinical studies involving over 6,000 obese subjects conducted in the Americas and Europe. In the United States, the FDA requires two years of safety data before approving anti-obesity medicines. The pharmaceutical company conducted those trials as part of its patent application process. As with many artificially produced drugs, adverse events proliferated. A meta-analysis by Christensen et al. (2007) published in the Lancet of Rimonabant safety data revealed an increased risk for suicidal ideation in patients, and two suicides across the two-year Rimonabant clinical trial program were recorded.

Furthermore, an analysis of data collected from four double-blind, randomized controlled trials by Buggy et al. (2011) demonstrated that 20 mg per day of this synthetic cannabinoid increased the risk of psychiatrically adverse events, specifically depression and anxiety. The conclusion of the FDA meta-analysis of Rimonabant safety data revealed an increased risk for suicidal ideation in patients. These findings resulted in the FDA withdrawing marketing authorization for Rimonabant because the adverse psychological effects could not be addressed. These results beg two questions. First, what was the mechanism causing these emotional disorders? Second, would a botanic cannabinoid CB1 receptor antagonist produce similar results?

An individual analysis of these questions brought up others and indicates how little we know about similarities and differences between synthetic cannabinoid medicines and botanic cannabinoid supplements. Reviewing the literature comprehensively also showed how little we understand about Rimonabant. Anavi-Goffer et al. (2012) identified two possible phytocannabinoid equivalents to Rimonabant. Both have significant implications concerning the proper regulation of the endocannabinoid system in relation to obesity, diabetes, and depression. One hundred sixty studies in the Medline Complete database published in the last five years classify the synthetic cannabinoid as an antagonist CB1 at the CB1 receptor, while 11 studies identify it as an inverse agonist. Scientific truth is not determined democratically, and the manufacturer of Rimonabant stopped answering questions about the synthetic cannabinoid in 2009 with the claim that the information was proprietary. This distinction between antagonist and inverse agonist is critical because it speaks to the mechanism causing the adverse reaction. McPartland et al. (2015) described a CB1 receptor inverse agonist as a receptor blocker, precluding the attachment of anandamide, the body's natural antidepressant endocannabinoid. Theoretically, blocking the attachment of anandamide to the CB1 receptor could conceivably result in depression. However, if merely blocking the CB1 receptor is enough to produce depression, any inverse agonist that blocks that receptor would prohibit the attachment of anandamide. If the eleven studies are correct, and Rimonabant merely acts as a receptor blocker, its phytocannabinoid equivalent would be cannabidiol (CBD).

De Petrocellis et al. (2012) demonstrated that CBD provides an astonishing benefit for hyperglycemia through its anti-inflammatory and antioxidant properties while modulating the cardiovascular stress response. However, because of the half-century prohibition of research on botanic cannabinoids in the United States, this has never adequately been studied. Given the proliferation of companies legally marketing CBD isolate botanic cannabinoid supplements throughout the nation, studies examining the potentially deleterious effects of these products on a human population are arguably necessary. According to data from the National Conference of State Legislators, in 2018, only four states banned access to medicinal botanic CBD; currently, that number is one (Nebraska). Cannabinoids derived from hemp are legally marketed in the remaining states as a treatment for various autoimmune disorders despite the long-held DEA and FDA contention that synthetic cannabinoids are medicinal and botanic cannabinoids are among the most addictive and dangerous molecules humans can ingest. With 99% of the nation allowing therapeutic CBD use, clinical and policy concerns regarding the mental health effects of botanic cannabidiol intromission should arguably be examined.

The National Institute of Drug Abuse subsidizes studies designed to prove the deleterious effects of botanic cannabinoids (Russell et al., 2019) while blocking inquiry into their potential benefits. If Rimonabant is an antagonist at the CB1 receptor, the depressive mechanism is different, but a study would fit into NIDA's subsidization wheelhouse. From a biomolecular perspective, one of the most important studies a quality supply of botanic cannabinoids could be used for is an analysis of the mechanism by which CB1 antagonism might cause depression. Theoretically, depression could be caused by the synthetic nature of Rimonabant, the byproduct of the antagonism of the CB1 directly, or the result of blocking the CB1 receptor, thereby prohibiting the binding of anandamide, the body's natural antidepressant endocannabinoid.

A study by Ahn et al. (2012) indicated that most of the central nervous system actions of cannabinoids, whether they be botanic, endogenous, or synthetic, are related to an affinity for binding with the CB1 receptor. If Rimonabant is an antagonist at the CB1, the phytocannabinoid equivalent has been determined by Di Marzo (2008) and confirmed by Tudge et al. (2014) to be Δ9-Tetrahydrocannabivarin (THCV). THCV appears to be anomalous in the world of botanic cannabinoids as the only known phytocannabinoid antagonist at the CB1 receptor. Still, Karch et al. (2007) described the chemistry of over 100 botanic cannabinoids, but because of the ban on their research, information is lacking about how most act on the body's various receptors. What we do know has to do with the "nature of science." In scientific research, it is generally the anomalies that end up being important. Dawson (2018) considered this particularly significant in devising a Complimentary Alternative Medicine (CAM) approach utilizing biomolecular psychology principles in treating obesity and diabetes.

Biomolecular psychology also provides insights into the development of CAM approaches targeting the endocannabinoid system in treating addiction disorders. Addiction disorders continue to take a toll on public health in most industrialized countries. According to the National Institute of Drug Abuse (NIDA, 2020a), more people die each year due to the ingestion of prescription medications and legal substances than by illegal ones. Data accumulated by NIDA (2020b) further reveals that millions of people die from addiction to tobacco and alcohol addiction every year, and currently accepted pharmaceutical treatments for addiction disorders demonstrate limited efficacy and application. Nutt et al. (2015) suggested that for these reasons, there is a need to understand better the brain's biomolecular mechanisms in developing and applying new and better-targeted interventions.

Dawson and Persad (2022) provided historical context illustrating the psychosocial, political, and bureaucratic barriers to applying biomolecular approaches to substance use disorders, focusing on what are "arguably the most stigmatized molecules in America." (p. 64). They delineated a biomolecular treatment strategy designed to mitigate multiple types of addiction disorders by influencing the dopamine and serotonin neurotransmitters' activity through phytocannabinoid supplementation of the endocannabinoid system and proposed a strategy for circumventing bureaucratic obstacles.

Humanity has struggled to understand the nature of drug abuse and addiction for centuries. There have been multiple false starts and a few successes throughout the years. Despite the traditional resistance to the medical use of botanic cannabinoids, their effectiveness in treating addictive outcomes is beginning to be understood from a biomolecular perspective. Psychoactive substances have been used since the earliest human civilizations, but biomolecular-based theories of why addiction to these substances occurs were not proposed until the scientific renaissance. The etiology of addiction is complex and reflected in the reoccurring pendulum swings between opposing attitudes concerning the causes of addiction that are still being debated. Crocq (2007) described views ranging from whether addiction should be considered a sin or a disease and whether the reason is moral or medical. Consensus has not been achieved within the scientific community concerning whether the drug, individual vulnerability, or social influences cause addiction. Baum (2016) even debates whether these substances should be regulated or freely available.

Katcher (1993) depicted Dr. Benjamin Rush, a signatory of the Declaration of Independence and founder of the first medical school in the United States, arguing in 1784 that alcoholism was a disease. Still, he possessed few scientific resources to study the problem. The intricacies of a biomolecular response to a substance would not be understood until the instrumentation was developed to measure a biochemical reaction and integrate this knowledge with complex cellular biochemistry. This instrumentation would not be invented until the 1980s (Borch & Rantala, 2015), so psychosocial and cultural explanations of addiction predominated. By the late 1700s, unrestrained alcohol use had become a significant public health problem for a substantial number of Americans. Rush published and frequently presented (Rush, 1814) and (Rush, 1816), correcting erroneous notions about alcohol's presumed medicinal benefits and accurately describing more than a dozen alcohol-related health problems. His speeches and publications launched the beginning of the American temperance movement.

Parsons and Hurd (2015) recounted the development and utilization of microdialysis instrumentation in the late 1980s and early 1990s. This technology significantly enhanced our understanding of the endocannabinoid system, a biological system composed of endogenous lipid-based retrograde neurotransmitters (endocannabinoids) that bind to cannabinoid receptors and cannabinoid receptor proteins expressed throughout the vertebrate central and peripheral nervous systems. They described this system as critical in regulating the physiological processes that underlie addiction disorders.

The formalized prohibition on the research of botanic cannabinoid molecules in the United States lasted 47 years and ended on December 21, 2018, when President Donald Trump ratified the Farm Bill (Smith, 2019). This bill legally reclassified botanic cannabinoids extracted from *Cannabis sativa* composed of less than 0.3% THC (hemp) as an agricultural product rather than a controlled substance, thereby legally (not scientifically) differentiating them from the molecules produced by cannabis varieties with higher THC content. When Donald Trump signed this bill, interstate transportation of the 151 known phytocannabinoid molecules became permitted, provided they originated from *Cannabis sativa* institutionally classified as hemp.

While the prohibition of research on botanic cannabinoid molecules has been inadvertently lifted, the paradigm to which cannabinoid scientists have been institutionally mandated to adhere remains. This paradigm was established in 1906 when the emergence of the FDA from the U. S. Patent Office coincided with the industrialization of synthesized medicines.

Janssen (1981) describes the current paradigm of medicine under which the science of medicine currently operates as originating in 1906 and coinciding with the emergence of the U.S. Food and Drug Administration (FDA) from the U. S. Patent Office and the industrialization of synthesized medicines. It was then that the FDA stated that by institutional mandate, it would only evaluate the efficacy of purported medicinal molecules if a patent accompanied the application for evaluation. Since naturally produced molecules cannot be patented, these nutraceutical molecules were institutionally eliminated from gaining the glorifying label of FDA-approved (Janssen, 1981). Thus, the medicinal molecules that the science of medicine relied on for treating disease and mood disorders for 12,500 years were eliminated from the arsenal of war against diseases and mood disorders on which humanity relied. This has resulted in a flawed paradigm of medicine in which the molecules that nature altruistically provides for the treatment of disease and mood disorders were eliminated from humanity's arsenal in its war against disease and mood disorders. The next century became devoted to denouncing these molecules and the individuals who used them (Herer, 1993). This stigmatization culminated in the early 1970s with the Nixon mandate that only the medicinal potential of synthetic cannabinoids was allowed to occur or be researched within the confines of the United States (Patton, 2020). Thus, a scientific paradigm psychosocially emerged over the course of a century, which viewed synthetic cannabinoids as medicinal, and the naturally derived cannabinoids with which humanity engaged in its war against disease and mood disorders for 12,500 years were institutionally eliminated from humanity's medicinal arsenal.

Summary

Since the Neolithic era, the science of medicine has been engaged in a war against disease and mood disorders, and until the early 1900s, botanic cannabinoids and other nutraceutical molecules were the only bullets utilized by humanity in this war. By the 1900s, the science of medicine had become based on the paradigm that these molecules were effective, and cannabis extractions were the most prescribed medicines by physicians at the beginning of the 20th century. In 1906, the Food and Drug Administration emerged from the Patent Office with an institutional decision that they would only evaluate for approval purported medicinal molecules if a patent accompanied them. Synthetic molecules can be patented. Natural molecules cannot.

Throughout the next century, the worldview under which the American Healthcare Delivery System operates entailed the adoption of the paradigm that only synthetic cannabinoids may be considered medicinal because only these molecules can achieve the glorifying FDA-approved label.

This paradigm was reinforced over five decades ago by the Nixon mandate prohibiting research of biologically produced cannabinoids and is currently in what Thomas Kuhn (1962) termed a state of crisis regarding the potential therapeutic properties of botanic cannabinoids. The dominant paradigm established and decreed by the mission statements of the DOJ, DEA, FDA, NIH, and NIDA demands acceptance of the paradigm that synthetic cannabinoids possess medicinal properties (Stone & Robert, 2022) and botanic cannabinoids are dangerous, having none. To ensure their economic security, cannabinoid scientists were institutionally mandated and economically incentivized to accept and promote this worldview for more than half a century. Until the passage of the Farm Act, studies that might challenge this paradigm were deemed illegal. This five-decade period defined the science of cannabinoids and dictated the methods of solving puzzles that arose. Ironically, this period was a phase that (Kuhn, 1962) would have referred to as "normal" science. The established paradigm dictated how observational data was perceived, experiments were designed, and the results were interpreted.

With the methods established and the assumptions defined, this paradigm flourished. However, deaths and other adverse events resulting from synthetic cannabinoid use and an accumulation of anomalies have challenged the dominant paradigm. Thomas Kuhn (1962) advanced the notion that the scientist's role is to design studies with the possibility of producing results that challenge the dominant paradigm. He coined the word "revolution" to describe dramatic changes in scientific worldviews. Revolutionary science is torturous and painful because it shakes all confidence that science has in its present theories and underlying paradigms. Paradigm shifts occur gradually when the dominant paradigm is termed to be in "crisis."

With changes in state regulations and the Farm Act's implementation, exploring the dominant paradigm's limitations is now possible. This is the nature of science. When too many anomalies appear that current theories cannot explain, a period of "crisis" results and political and economic events fuel the search for new understandings. This is the stage we are in concerning botanic cannabinoid-based medicines. Studies are only now beginning to be proposed that might subvert the accepted assumptions. Dawson (2019a) observed that history has demonstrated that whether the opposition to attaining and disseminating scientific knowledge is politically or religiously motivated, humanity has the potential to ensure this knowledge is eventually acquired.

Flawed Paradigms and their Influence on our Understanding of Addiction

The historical analysis described above illustrates the convoluted nature of the construct of illicit substances and the political, cultural, psychosocial, and bureaucratic influences that brought about this convolution. The paradigm of the nature of addiction has changed, and our understanding of the mechanism that causes someone to become addicted to a substance has evolved. Pickard et al. (2015) observed that for much of the twentieth century, theories of addictive behavior and motivation became polarized between the moral and medical models of addiction.

Carson et al. (2016) described the moral model as the easiest to understand and most traditional approach to addiction. The model focused on the idea that addiction is indicative of an individual's moral failures, and the legal system should be employed to punish illicit substance users for their indecency. Addiction in this model is considered a result of an individual's poor life choices. Despite the propaganda promoting a subsequently disproved economic justification of this model by politicians and leaders in the law enforcement industry, the moral model has fallen into disrepute in academia. On June 19, 2017, analysts for Pew Charitable Trust submitted a letter to the President's Commission on Combatting Drug Addiction and the Opiate Crisis (Stein, 2017), outlining an analysis of whether state drug imprisonment rates are linked to the nature and extent of state drug problems. This was a significant question as the nation faced an intensifying opioid epidemic. Pew compared publicly available data from law enforcement, corrections, and health agencies from all fifty states. The analysis found that no statistically significant relationship exists, indicating that incarcerating drug addicts is ineffective in the nation's war against drugs. If incarcerating addicts were effective, states would experience reduced drug abuse rates. Instead, they found that higher imprisonment rates did not correlate with lower drug abuse rates, drug arrests, or drug fatalities. These findings added to mounting evidence (Report to Congress, 2012) demonstrating that incarceration was not a viable solution for addiction. Still, in 2018, the most significant politician in America promoted the Moral Model of Addiction. On Monday, March 19, 2018, President Donald Trump called for harsher prison sentences when he announced his program to combat the opioid epidemic (Lopez, 2020).

The Medical Model, which views the construct as a disease, is diametrically opposed to the Moral Model of Addiction. The Medical Model considers addiction analogous to any other illness and the predisposition to contracting an addiction disorder outside the individual's control. Animal studies (Elmer et al., 2010; Freet et al., 2013; Griffin et al., 2007; Le Foll et al., 2009) supported this position. These studies utilized C57 genotype mice in addiction experiments because their genetic makeup makes them more sensitive to the rewarding effects of morphine, cocaine, and alcohol than other genotypes in the same way genetic factors predispose persons to addiction.

The Medical Model proclaims addiction is a disease, not a sin, as claimed for over two centuries, beginning in the early 1800s. Alcoholism was declared a disease by the American Medical Association in 1956

(Schneider, 1978), and the organization classified other addictions as diseases in 1989. While the American Medical Association is an established institutional authority, if politicians who make medical decisions for the population accepted addiction as a disease, this assertion would not need to be continuously reasserted. In the context of addiction, "disease" is frequently viewed entirely differently from the word "disease" used for any other condition. It has not been necessary for the American Medical Association to repeatedly declare that AIDS or diabetes are diseases. While a segment of the population and certain politicians became convinced that COVID-19 was a hoax, no one has ever asserted that it is not a disease.

Despite the American Medical Association's repeated proclamations, many people do not accept addiction as a legitimate medical condition. A 2010 study by Pescosolido et al. (2010) examined Americans' perceptions about alcoholism, finding that 65 percent of 630 people polled in a general population sample felt the condition was due to "bad character," while 47 percent viewed it as a problem of brain chemistry or genetics. Four years later, a survey study by Barry and McGinty (2014) found a robust moral stigma associated with addiction. Seventy-eight percent of responding participants stated they did not want to work with someone with an addiction disorder, and ninety percent said they would not want someone with an addiction to marry into their family. These rates are significantly higher than those who would similarly reject someone with diabetes.

The stigmatization is ongoing. If addiction is truly a disease, it is the only one that can get the sufferer arrested simply for displaying symptoms. Punitive incarceration for being addicted undermines the disease model. Images of arrests and police officers standing in front of massive quantities of drugs promote the idea that drug use is criminal and undermines the medicalization model, with drug users depicted as sinful by nature. The ACLU (2021) identified addiction as the only existing disease for which a judge can deny medical treatment. While advances in neuroscience provided compelling evidence to support a medical perspective of problematic substance use and addiction, the science is still in its preliminary stages (Buchman et al., 2010), and theories about how addiction emerges have been neither universally accepted nor wholly understood.

Crow (1972) described a significant event occurring in biomolecular psychology in the 1970s when a mechanism of addiction was identified. It was found that rats would repeatedly and willingly electrically self-stimulate areas in the brain, which were subsequently demonstrated to comprise a set of dopamine neurons. The Crow (1972) study explained the results of an earlier experiment by Stein (1964), indicating that stimulants enhanced this neurotransmitter's actions. A subsequent series of in vivo experiments by Wise & Bozarth (1987) and Robinson and Berridge (1993) demonstrated that blocking dopamine receptors with neuroleptic drugs impaired the reinforcing effects of addictive drugs in rats and primates. The amalgamation of these studies placed dopamine as the central neurotransmitter in addiction and indicated that it played a critical role in reward, motivation, and incentive behavior.

The next conceptual breakthrough coincided with the development of microdialysis sampling techniques pioneered by a group of researchers in Sardinia, Italy. Microdialysis sampling by Di Chiara and Imperato (1988) produced conclusive evidence that drugs of abuse release dopamine in the basal forebrain, a preoptic area of the hypothalamus known as the nucleus accumbens. This resulted in a general theory of addiction in which addictive drugs release dopamine, but non-addictive substances do not. From this point, the field developed rapidly, with replications of the early animal findings of dopamine being released by 'addictive' drugs and confirmations in humans by Volkow et al. (2003) using neurochemical imaging. These studies' results led to immense investment in research to alter dopamine neurotransmitter function to treat addiction. Positron imaging tomography (Volkow et al., 1994) and single-photon emission computed tomography (Lareulle et al., 1996) provided critical breakthroughs in our understanding of the human dopamine system and its role in addiction when it was demonstrated that these technological innovations could be used to measure dopamine release in the human striatum. It was later demonstrated by Volkow et al. (1999) that the magnitude of this increase could predict the euphoria or 'high' produced by a drug. This proved that in humans, the feeling of pleasure produced by addictive drugs is mediated by striatal dopamine release by the same mechanism as in animals (Di Chiara & Imperato, 1988), and addiction has come to be viewed as a disorder of the dopamine neurotransmitter system (Nutt et al., 2015).

The dopamine theory of addiction has generated acceptance by biomolecular psychologists because drugs that induce dopamine release repeatedly correlate with feelings of pleasure or euphoria. This sensation of joy or bliss is considered indicative of psychoactivity. According to the moral model of addiction, psychoactivity should only be induced by legal chemicals like alcohol, tobacco, caffeine, or physician-prescribed pharmaceutical medications.

The development of the technology capable of analyzing neurotransmitters and applying the results to the dopamine theory of addiction profoundly affected the creation of synthetic drugs designed to target the brain. Pharmaceutical companies used ventral striatal dopamine release assays to estimate the abuse potential of new medicines (Uguen et al., 2013), rejecting compounds if they were determined to be psychoactive, as determined by increased dopamine levels. It might be argued that this was a concession to the moral model of addiction, that anything that results in a pleasurable sensation should be illegal. Still, this view is disconcerting because human (Brown & Gershon, 1993; Robinson et al., 2012) and animal studies (Roth-Deri et al., 2009; Tye et al., 2013) have conclusively demonstrated that dopamine activity in the ventral striatum is critical in resistance to depression.

As already implied, the moral model of addiction has produced barriers to cultural and political acceptance of botanic cannabinoid medicines. Kenny and Zito (2022) described how this model affected where botanic cannabinoids would fall when the Controlled Substances Act was created. The United States Drug Enforcement Administration (DEA) classifies chemicals, drugs, and certain substances used to make drugs into five categories or schedules depending upon the abuse or dependency potential and their acceptability as medicine. The potential for abuse is the determining factor in scheduling a drug (Kenny & Zito, 2022), with Schedule I drugs having a high potential for abuse and creating severe psychological and physical dependence. The lower the substance appears on the Schedule, the less potential the drug has to be abused. Legal drugs like alcohol and tobacco do not appear on the Schedule (Robbins, 2018), meaning the DEA considers them outside their purview as the paramilitary entity tasked with adjusting the Schedule and enforcing it upon the citizenry.

Unable to provide the studies the DEA claimed to have that demonstrate the abuse potential of biological cannabinoids, the justification for them to remain Schedule I became based on claims that a particular phytocannabinoid is psychoactive and, therefore, pleasure-inducing. A compound's psychoactive nature is considered indicative of its potential for abuse. In 1971, the United States declared a 'War on Drugs,' targeting the entirety of the known botanic cannabinoids. Research of biologically produced cannabinoids was officially prohibited in 1972, and research universities were mandated to forbid biological cannabinoid studies and were warned that conducting such research could jeopardize their federal funding. At the time, the only known biological cannabinoids were botanic, the cannabinoids derived from the cannabis plant. These non-synthetic molecules were classified as Schedule I drugs, the category reserved for the most dangerous substances humans can ingest. This classification for the biological cannabinoids worked well for the next two decades because it fit perfectly into the moral model of addiction, which fueled the drug war. Dawson and Persad (2022) recounted the DEA flourishing during this time through asset forfeitures and massive budgeting allocations provided by Congress to ensure the Judicial System punished Americans for possessing cannabinoid molecules produced biologically.

Things became problematic for the law enforcement agency in 1992 when the United States funded researchers at Hebrew University in Jerusalem (Devane et al., 1992). This group identified the first endogenous biological cannabinoid (endocannabinoid) and named it anandamide. Shortly thereafter, the National Institute of Drug Abuse (NIDA, 2021) declared the endogenous cannabinoid anandamide and the botanic cannabinoid Δ9-Tetrahydrocannabinol (THC) to be equivalent molecules and, along with neurotransmitter action evidence, justified this position with the image provided below. Suddenly, it could reasonably be argued that all Americans were in possession of an illicit substance merely by being alive (Dawson & Persad, 2022).

Endocannabinoid/Phytocannabinoid Equivalents

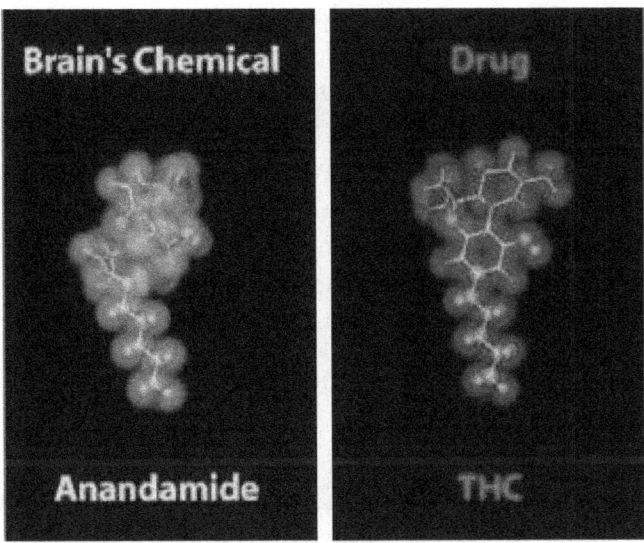

Courtesy of NIDA

Note. THC's chemical structure is similar to the brain's chemical anandamide. The similarity in structure allows drugs to be recognized by the body and to alter normal brain communication (NIDA, 2021).

Cannabidiol was determined by the United States Department of Health and Human Services (2003) as the phytocannabinoid equivalent to the endocannabinoid 2-arachidonoylglycerol (2-AG) because each acts on the body's receptors in the same ways (Dawson, 2018), and other biological cannabinoid equivalents quickly followed. Hill & Newlin (2002) recounted the refusal of the DEA to deschedule botanic cannabinoids and recategorize them as unscheduled substances like alcohol and tobacco. Instead, politicians admitted the war against the known botanic cannabinoids was lost and shifted the attack to Δ9-THC by misappropriating the term psychoactive. Δ9-THC would remain the botanic cannabinoid with the most destructive potential due to its purported psychoactive nature.

Sexton et al. (2016) described the psychoactive property coming from its ability to release dopamine by activating the CB1 receptor the same way anandamide does, thereby providing a sense of pleasure abhorrent to the moral model of addiction. In a classic bit of hypocrisy, at the same time the United States was claiming cannabidiol had no medicinal properties and great potential for abuse, it filed patent number 6630507 (United States Department of Health and Human Services, 2003) claiming the rights to the use of a non-psychoactive phytocannabinoid to treat neurological conditions resulting from concussion or stroke, and its use in the treatment of Alzheimer's, Parkinson's, autoimmune diseases, in addition to HIV dementia. Cannabidiol was claimed to be non-psychoactive and, therefore, the non-controversial and 'safe' botanic cannabinoid.

Somehow, the patent survived the application process. The federal government now holds the license on a federally illegal, unpatentable, natural form of non-psychoactive cannabidiol to treat Alzheimer's disease, Parkinson's disease, human immunodeficiency virus dementia, Down's syndrome, and heart disease (United States Department of Health and Human Services, 2003). Within the patent application, it is repeatedly stated that the cannabidiol being patented has no psychoactive properties. One of the most significant differences between science and politics is that in science, terms must be defined. In the patent application, the term "non-psychoactive" is undefined. Thus, the standard definition must apply. The World Health Organization (1994, p. 53) provided this definition; "Any substance that affects mental processes when ingested is considered to be psychoactive."

A study by Oakes et al. (2019) conclusively demonstrated that cannabidiol acts as an allosteric modulator of serotonin levels, affecting mental processes by providing a sense of contentment. The study further revealed that the endogenous cannabinoid receptor agonist cannabidiol agonizes the 2-arachidonoylglycerol (2-AG) receptor, thereby enhancing serotonergic signaling by increasing endogenous serotonin concentrations. This study confirmed the results of Sartim et al. (2016), who used rat models to effectively demonstrate the involvement of serotonin receptors in the antidepressant-like effect of cannabidiol through infralimbic and prelimbic microinjections of the botanic cannabinoid into the ventral medial prefrontal cortex and submitting the subjects to forced swimming or to open field tests. The intromission of cannabidiol significantly reduced the immobility time in the forced swimming test without changing locomotor activity in the open field test, indicating the administration of CBD into the ventral medial prefrontal cortex induced antidepressant-like effects potentially through the indirect activation of the 5-HT1A receptors. The Sartim et al. (2016) study supported the results of Sales et al. (2018), who investigated the participation of serotonin in CBD-induced antidepressant-like effects in forced swimming tests and demonstrated that serotonin depletion impaired cannabidiol-induced behavioral effects. The results of this study indicated that the antidepressant-like effect induced by cannabidiol in the free-swimming test was dependent on serotonin concentrations in the central nervous system. Taken together, the results of these studies imply that if the government has patented a form of cannabidiol that does not alter serotonin levels when intromitted, it is probably unique in the cannabinoid world. Dawson (2019b) posited that it is more likely that the government has claimed a patent on a non-psychoactive form of cannabidiol, which cannot, in reality, exist.

While the United States holds a patent on a unique form of cannabidiol purported not to affect brain chemistry, research is progressing on existing cannabidiol compounds that activate the 5-H1A (serotonin) and TRPV-1 receptors and inverse-agonize the CB1 receptors, thereby contributing to the treatment of morphine addiction (Katsidoni et al., 2013), cocaine addiction (Luján et al., 2018), nicotine addiction (Morgan et al., 2013), heroin addiction (Qian et al., 2020), THC addiction (Shannon & Opila-Lehman, 2015), and methamphetamine addiction (Shen et al., 2022).

The dopamine theory of addiction purports addiction to be a disorder of the dopamine neurotransmitter system and is, by definition, a medical issue (Nutt et al., 2015). As with the array of endocannabinoid deficiency disorders, neurotransmitter disorders are not yet recognized as a disease. Although not officially a disease and, therefore, not technically a primary condition, Dawson and Persad (2022) cite evidence that neurotransmitter disorders can effectively be treated by nutraceutically manipulating the system that oversees the biomolecular mechanism responsible for the neurotransmitters' production. While addiction has come to be viewed as a disorder of the dopamine neurotransmitter system (Nutt et al., 2015), pharmaceutically targeting this neurotransmitter has not led to new treatments. Tanimura et al. (2010) demonstrated that while dopamine deficiency has been verified to have a significant role in the susceptibility to addiction, these effects are mitigated by the activation of the serotonin neurotransmitters.

As with many biomolecular psychology concepts, the simplest way of reaching the target is to modulate the endocannabinoid system. The cannabinoid receptors and their endogenous agonists and antagonists (endocannabinoids) are ubiquitously distributed throughout the central nervous system and play a critical role in regulating neurotransmitter production and excitability. A review of in vivo studies by Viveros et al. (2005) supported by an invitro study by Thomas et al. (2007) demonstrated that 2- arachidonoylglycerol (2-AG) plays an essential role in mitigating stress-related behaviors and the mood-elevating and euphoric effects resulting from the intromission of dopamine-enhancing drugs. Experiments by Devane et al. (1992) and Parsons and Hurd (2015) revealed that the two most critical endocannabinoids related to addiction are anandamide and 2-AG. Anandamide is the endocannabinoid that agonizes the CB1 receptors responsible for activating dopamine neurotransmitters (Thomas et al. (2007). This results in the sense of pleasure erroneously referred to as psychoactivity. The bliss resulting from dopamine activation has been demonstrated to be the biomolecular mechanism that makes psychoactive drugs addicting. 2AG is the endocannabinoid responsible for serotonin activation (Mechoulam et al., 1995), the neurotransmitter that provides a sense of happiness, contentment, and well-being. For these reasons, from a biomolecular perspective, addiction treatment is more likely to be effective by activating serotonin while repressing dopamine. The most useful phytocannabinoid for accomplishing this task is cannabidiol. CBD inverse agonizes the CB1 receptor (Müller & Homberg, 2015; Peng et al., 2022), blocking its access and thereby mitigating the effects of addictive drugs. Simultaneously, it activates the serotonin neurotransmitters (De Gregorio et al., 2019). This biomolecular mechanism explains how CBD's intromission effectively treats every substance abuse disorder studied.

Systematic reviews of multiple studies by Prud'homme and Cata (2015) and Siklos-Whillans et al. (2021) indicated that biomolecular manipulation of the endocannabinoid system through phytocannabinoid supplementation would complement the Matrix Model, an integrative therapeutic approach incorporating the most efficacious aspects of behaviorism, person-centered therapies, cognitive behavioral therapy, the twelve-step approach, and motivational interviewing. As a harm reduction strategy, phytocannabinoid supplementation is supported by evidence demonstrating its efficacy for pain relief and as a substitution for multiple illicit drugs, alcohol, tobacco, and pharmaceuticals. Animal models demonstrate that phytocannabinoids reduce withdrawal, which contributes to drug-seeking behavior (Hurd et al., 2015), but statistically significant human clinical trials are lacking. With few exceptions, animal trials involved the addictive drug being administered intravenously, followed by the animals receiving a CBD systemic injection. While this methodology is deemed acceptable for rats and mice because they are excluded from the Animal Welfare Act, such an approach might not pass IRB scrutiny for human clinical trials in many industrialized nations (Banerjee, 2015).

The FDA has rejected nearly all ingestion methods of botanic cannabinoids, but transdermal drug delivery of nicotine developed by the pharmaceutical industry has made an essential contribution to medicine and can easily be appropriated to deliver specific doses of botanic cannabinoids into the bloodstream through a porous membrane. Pastore et al. (2015) described transdermal patches providing steady and consistent permeation of a drug through the skin, leading to more constant plasma supplement levels, which is usually the goal of therapy.

The absence of peaks and troughs in plasma concentration levels results in significant improvement compared to traditional ingestion methods of nutraceuticals. Intromission through a transdermal patch has FDA approval, and CBD intromission does not result in dopamine activation, so the detestable pleasurable sensation is avoided. Still, serotonin levels are enhanced, providing the subject with a sense of contentment and well-being.

Allen (2016) described the endocannabinoid system as the single most important scientific medical discovery since the recognition of sterile surgical techniques. Alger (2013) expounded that as our understanding of its complexities expands, it will become increasingly apparent that the endocannabinoid system connects and controls all physiological and psychological processes. Even so, creating and conducting human clinical trials to test these hypotheses in the United States is problematic. A survey study conducted by Allen (2016) revealed that many American scientists are ignorant of even the endocannabinoid system's existence and would likely need to be trained if a clinical trial were to be conducted in America. Reid (2020) described the systemic institutionalization prohibiting physicians associated with hospitals from learning about or discussing the endocannabinoid system because doing so could jeopardize the hospital's federal funding. The historical timeline in this disquisition illustrates the breaches of ethics the United States government has committed to ensure knowledge of botanic cannabinoids is repressed. This suggests that the United States should be eliminated as a potential country in which to conduct human clinical trials to test the efficacy of treating addiction disorders by targeting the endocannabinoid system and that if a clinical trial of this nature is to occur, it might be more advantageous to conduct such a study in a more scientifically progressive country and one more conducive to innovative and non-synthetic approaches.

Chapter 3

The Institutionalized Stigmatization of Botanic Cannabinoid Medicines

The stigmatization of the attribute of medicinal botanic cannabinoid use is unique in that its entire stigmatization history is exceptionally well documented from its origination through its perpetuation and culmination and purported diminishment. Examining the origins and development of the institutionalized stigmatization of botanic cannabinoid use potentially provides a mode of comparison into the origins and development of other institutionally stigmatized attributes.

Friedman et al. (2022) described the institutionalized stigmatization of cannabis by politicians, business leaders, and other upper-echelon entities beginning with the inception of the Department of Justice and the Federal Bureau of Narcotics. The inception of these institutions served to condition the American psyche to stigmatize botanic cannabinoid medicines. Still, as eloquently explicated by Durkheim (1951), the origin of a phenomenon does not explain its persistence or its transformations. This chapter aims to examine in depth in depth the origin of the institutionalized stigmatization of therapeutic botanic cannabinoid medicines, its development, culmination, and purported diminishment through the lens of biomolecular psychology.

Through the integration of scores of scholarly articles about the historical stigmatization of phytocannabinoid-based medicines, a clear picture is painted of the psychosocial, institutional stigmatization conditioning that has occurred in American society since 1865. This picture also reveals the convoluted strategies upper-echelon entities utilized to originate and perpetuate this psychosocial stigmatization conditioning. A psychosocial historical analysis of the stigmatization of phytocannabinoid-based medicines demonstrates that this phenomenon is correlated with the establishment of the Department of Justice in the wake of the Civil War.

Shugerman (2014) and Harris and Martin (2021) described how Congress established the Department of Justice at the end of the Civil War, bestowing on it the primary purpose of enforcing reconstruction and protecting the civil rights of ex-slaves. However, the institutionalized stigmatization of individuals of color was already deeply embedded in American society, resulting in the Department of Justice running in a more realistic direction, achieving with its subsidiary organization, the Federal Bureau of Narcotics, the more attainable goals of increasing its size and power and enlarging the machinery of government. Dawson (2024), Hatzenbuehler (2016), and Salter et al. (2018) detailed how these two governmental institutions established, developed, and perpetuated the systemic structural stigmatization of botanic cannabinoid medicines that would last to the present day.

Roberts and Rizzo (2021) and Williams and Williams-Morris (2000) depicted discrimination as an established American tradition, one for which Americans have proudly fought and died. Congress ratified the Narcotic Drug Import and Export Act and created a subsidiary of the Department of Justice, the Federal Narcotics Control Board, in 1922 (Patton, 2020). The Department of Justice, through its subsidiary organization, the Federal Bureau of Narcotics, defined "narcotic" as "any drug used by individuals of low socioeconomic standing," and this definition has never been rescinded (Patton, 2020, p. 7). Throughout much of the next century, botanic cannabinoid use became associated with maligned and structurally stigmatized ethnic and racial minorities (Harris & Martin, 2021), and the stigmatization of the medicinal use of botanic cannabinoid use continued to manifest in innumerable forms from multiple institutional entities throughout the next century.

Extrapolating from the research of Janssen (1981) and Crocq (2020) reveals the stigmatization of therapeutic botanic cannabinoid use by the American Healthcare Delivery System as originating in 1906, with the industrialization of synthesized medicines coinciding with the emergence of the FDA from the U. S. Patent Office. If it is possible to pin it down to a year, 1906 is the year the United States made the institutionalized decision that its goal was to be a synthetic society. Humanity has been engaged in a war with disease and mood disorders since Neolithic times. From the onset of the Neolithic Era until the industrialization of synthesized medicines, the only bullets humans had in their arsenal to wage this war were the medicinal molecules that nature altruistically provided. For those 12,500 years, until the FDA emerged from the patent office, the science of medicine was dependent on the medicinal molecules derived from nature's processes. Cannabis/hemp contained more medicinal bullets than any other plant, so these molecules were naturally the most prescribed medicinal molecules by physicians in the early 1900s.

Euphemistically, these nutraceutical molecules were the only bullets humans held in their arsenal to wage battle against disease and mood disorders until the FDA emerged from the Patent

Office with the declaration that it would not evaluate any purported medicinal molecules unless a patent of the molecule accompanied the application for evaluation. Natural molecules cannot be patented, so the medicines humanity had been reliant on in its 12,500-year war against disease and mood disorders were eliminated from humanity's arsenal, and the United States as an institution devoted the next century to stigmatizing these molecules.

Anguelov (2018) described the structural stigmatization of therapeutic cannabinoid use taking over a century and undergoing wildly peculiar convolutions, each having to do with upper-echelon influencers of society with economic or political incentives to destroy the species *Cannabis sativa* in the United States of America since colonial times. Despite the historical stigmatization of botanic cannabinoid medicines in the United States, attitudes about phytocannabinoid-based treatments have been dynamic, and policies related to the species *Cannabis sativ*a and its molecular components have undergone multiple phases and convolutions all originating from upper-echelon administrative entities that influenced the psychosocial structure and behavior patterns of American society.

Clark (1929) identified the first American law related to *Cannabis sativa* being enacted by British rule in Jamestown Colony, Virginia, in 1619, mandating that all colony farmers cultivate cannabis hemp, bestowing on them the much-sought-after prize of British citizenship for those who complied and levying fines on those who refused. Additional must-grow cannabis hemp cultivation laws were enacted in Massachusetts in 1631, Connecticut in 1632, and the Chesapeake Colonies throughout the mid-1700s. Cannabis hemp was legal tender throughout much of America from 1631 until the early 1800s, and Americans could pay their taxes with cannabis hemp for more than 200 years (Herndon, 1963). Throughout colonial America, the cultivation of cannabis hemp was mandated by the Crown, the plant being a necessary crop because its stalks and fibers were inexpensive and essential for the production of cloth, food, thread, oil, paper, and virtually all the rigging, anchor ropes, cargo nets, flags, shrouds of ships, and the tar produced from the stalks of Cannabis sativa (oakum) was critical for the protection of vessels against salt water.

From the early 1600s to the first part of the 20th century, cannabis was cultivated throughout the United States. McPartland (2017) described two forms of *Cannabis sativa* being grown, one for fiber and industrial purposes (now commonly referred to as hemp) and one for producing medicines. Botanic cannabinoid medicines found their way into the American pharmacopeia in 1851. However, an understanding of the biomolecular processes responsible for their efficacy was ill-defined and would not be explained until Devane et al. (1992).

Research by Herer (1993) revealed that by the end of the 19th century, *Cannabis sativa* extracts were the most widely prescribed medicines in the United States. Institutionalized stigmatization conditioning of the medicinal use of *Cannabis sativa* in America originated in the late 1800s (Hatzenbuehler, 2016), and as with many instances of institutional stigmatization conditioning, multiple upper-echelon entities were involved in its origination and perpetuation. These upper-echelon entities created or influenced the country's administration to establish and enforce policies affecting societal-level conditions and cultural norms to hinder targeted individuals' prospects, resources, and well-being. They played a critical role in the psychosocial stigmatization conditioning of medicinal botanic cannabinoid use in American society since the end of the Civil War.

Through the integration of scores of scholarly articles about the historical stigmatization of phytocannabinoid-based medicines, a clear picture is painted of the psychosocial, institutionalized stigmatization conditioning that has occurred in American society since 1865. This picture also reveals the convoluted strategies upper-echelon entities used to originate and perpetuate this psychosocial stigmatization conditioning. A historical analysis of the stigmatization of phytocannabinoid-based medicines demonstrated that this phenomenon is correlated with the industrialization of synthesized medicines coinciding with the 1906 emergence of the FDA from the Patent Office and the establishment of the Department of Justice and the Federal Bureau of Narcotics in 1870 and 1930, respectively.

From a psychosocial perspective, the man most responsible for America's institutionalized stigmatization of *Cannabis sativa* and medicinal botanic cannabinoid use was Harry Anslinger. Anslinger was appointed the first Federal Bureau of Narcotics commissioner in 1930 by his uncle and Secretary of Treasury Andrew Mellon (McWilliams, 1989), and he would hold this position for the next 31 years. Dawson and Persad (2022) asserted that Anslinger's most notable achievement was establishing the precedent that in the United States, politicians are appointed to make medical decisions for the people they govern. Anslinger was the first commissioner of the Federal Bureau of Narcotics, which laid the groundwork for the modern-day DEA, and he was the first architect of the war on drugs. Anslinger was appointed to the bureau in 1930, just as the prohibition of alcohol was beginning to crumble. From the moment he took charge, Harry was aware of the weakness of his new position. Anslinger needed to be able to justify his new bureau's existence financially. He knew he couldn't keep an entire department alive on narcotics alone. Cocaine and heroin were simply not used enough to sustain a whole bureau. At the time, very few people were using heroin and cocaine (Musto, 1999).

To fund this newly established bureau, Anslinger made it his mission to rid the United States of all drugs except alcohol and tobacco. He collaborated with other upper-echelon influencers of societal mores, such as the Dupont family and William Randolph Hearst. All had economic incentives and the upper-echelon positions to influence administrative policies and collaborated to initiate a disinformation campaign targeting cannabis hemp and, as a by-product of this campaign, stigmatize phytocannabinoid-based medicines. Anslinger's economic incentive has already been mentioned. He needed to be able to justify his bureau's existence. He could not sustain the Federal Bureau of Narcotics by targeting cocaine and heroin alone, as these were not used enough in the 1900s to financially justify the existence of an entire bureau (Musto, 1999). To fund his newly established bureau, Anslinger made it his mission to rid the US of all drugs except alcohol and tobacco. This included cannabis, which he chose to pursue with a vengeance.

He collaborated with other upper-echelon influencers of societal mores, such as the Dupont family and William Randolph Hearst. All had economic incentives and the upper-echelon positions to influence administrative policies and collaborated to initiate a disinformation campaign targeting cannabis hemp and, as a by-product of this campaign, stigmatize phytocannabinoid-based medicines. Anslinger's economic incentive has already been mentioned. He needed to be able to justify his bureau's existence. The Dupont Organization understood chemistry (Dwyer, 1997) and had the financial motivation to make America into a synthetic-based society. Nutraceutical phytocannabinoid-based medicines, hemp textiles, and industrial products represented competition requiring elimination. Research by Gerber (2004) revealed that in the 1930s, the paper-rich and timber tycoon William Randolph Hearst was also financially threatened by cannabis hemp, as it was poised to replace timber in manufacturing paper and, as such, a competitor requiring eradication.

Hearst, Dupont, and Anslinger conspired to create economic competition, and the media manipulation they engaged in to stigmatize cannabis hemp and the medicines *Cannabis sativa* provided can only be described as inspired. Garza (2018) depicted this trio operating at the upper echelon of society to influence information disseminated to administrators and the public by manipulating word choice and coining the word 'marihuana,' thereby playing on the societal structural stigmatization of Mexicans, which Hearst, through his publishing empire had been integral in perpetuating. Through word choice manipulation, this gang of three was able to construct a reality where cannabis hemp did not exist, and marihuana was a new threat. Referring to cannabis with a different name marked an acute shift in the American perspective of the supplement. Solomon (2020) details fearmongering in media outlets manipulating the zeitgeist of the American population by equating marihuana use to alleged crimes committed by Mexican immigrants. This criminal activity was frequently fabricated, exaggerated, and portrayed largely inaccurately by Hearst and the others, who did an admirable job constructing their narrative. The words hemp or cannabis were never used in the smear campaign decimated through the massive Hearst publishing empire. The Hearst media empire published stories claiming marihuana smoking Mexicans and African Americans would rape and disrespect whites. By 1936, Hearst had fabricated hundreds of horror stories (Nicholas

& Churchill, 2012; Solomon, 2020) associating marihuana use with violent crime, which Anslinger preserved in a 'Gore File' to refer to as the need arose.

Here (1963) described hemp farmers reading these stories and, because the word marihuana was used, never realized that their livelihood was being targeted, as the word had never been associated with the plant they cultivated. Anslinger soon began utilizing other forms of media manipulation. Using General Social Survey data in a historical analysis examining the relationship between media exposure and attitudes toward cannabis legalization, Stinger and Maggard (2016) compared the media coverage in the years 1975 to 2012 to the negative horror stories about marihuana depicted by the media throughout the 1930s Reefer Madness era when the plant was first outlawed. With the creation and exploitation of movies vilifying violence-inducing properties of marihuana Anslinger and Hearst invented, including Assassin of Youth, Marihuana, the Weed with Roots in Hell, and Tell Your Children (later renamed Reefer Madness), the media played a critical role in the origination and perpetuation of botanic cannabinoid use stigmatization. Reefer Madness began with this stigmatizing statement about marihuana, "There is a new drug menace destroying the youth of America. It is a violent narcotic" (Luginbuhl, 2001, p. 3).

Marihuana use triggering addiction, insanity, and violent crime is the Criminality Theory (Bonnie, 1974) and became the rationale for cannabis hemp prohibition. Data compiled in this study revealed that Americans' attitudes toward marijuana have evolved dramatically from the time it was criminalized in the 1930s through the present day, and public opinion favoring the legalization of cannabis has steadily risen since 1990. It is generally well-accepted that the media as an institution played a critical role in manipulating not only laws but also the general public's attitudes toward the cannabis plant.

Gelders et al. (2009) proposed that because many people in the United States have little direct knowledge of illicit drugs, they tend to get their information from the most common, easily accessed source, the mass media. Goode (1993) depicted the media as portraying worst-case scenarios, placing negative spins on events, and exaggerating the issues as a marketing strategy. Research by Stinger and Maggard (2016) revealed the initial panic over marihuana in the 1930s to be socially constructed by the media and government influence on public opinion. For example, in the 1930s, as the Federal Bureau of Narcotics Commissioner, Anslinger used the media to advance the structural stigmatization of botanic cannabinoid use by testifying to Congress in the 1937 Marihuana Tax Hearings. He testified, "Marihuana is the most violence-causing drug in the history of mankind" (Abrams & Guzman, 2015, p. 1). He then introduced his 'Gore file,' culled almost entirely from Hearst yellow journalism tabloids, including stories of ax murder, where the perpetrator reportedly smoked a joint four days before committing the crime. He continued in this vein, stating to Congress that 50% of violent crimes committed in the U.S. were perpetrated by Spaniards, Mexican Americans, Latin Americans, Filipinos, African Americans, and Greeks, and these violent crimes could all be traced directly to the use of marihuana.

Anslinger's influence played a significant role in introducing and passing the Marihuana Tax Act of 1937, which outlawed the possession or sale of marihuana without a tax stamp, which he ensured would be virtually impossible to obtain. In the Congressional hearing, Anslinger repeatedly claimed that marihuana use caused psychosis and, eventually, insanity and death. He stated that young people are 'slaves to this narcotic, continuing addiction until they deteriorate mentally, become insane, turn to violent crime and eventually murder.' The problem was that he had no scientific evidence to support this. He had contacted 30 scientists, and 29 told him cannabis was not a dangerous drug. He hired the 30th, Dr. James Munch, as the US government's official expert' on marijuana. In this position, Dr. Munch was responsible for presenting the scientific evidence to Congress that demonstrated marihuana intromission caused psychosis and, eventually, insanity.

As recounted by Bonnie and Whitebread (1974) in a 1995 speech to the California Judges Association annual conference by Charles Whitebread, Professor of Law at USC Law School, entitled *The History of the Non-Medical Use of Drugs in the United States,* Dr. Munch stated in the Congressional hearing that he had conducted a series of experiments with cannabis on dogs and had, for scientific purposes, tried marihuana himself. When asked how it affected him, Dr. Munch testified, under oath, that, after two inhalations of a marihuana cigarette, he was turned into a bat, flew around the room, and down a 200-foot-deep inkwell.

Congress had never heard of marihuana and was unaware of its relationship with its hemp-cultivating constituents. But the hour was late, and it was time to move on. In a vote they didn't bother to record (Holfield, 2013), on a matter of little interest, to protect the nation from the threat of this violent narcotic, a handful of Congressmen ratified a bill that would one day fill the nation's prisons to the brims. During this act of demonization, Anslinger continuously cited his Gore file, the accumulated body of misinformation published by Hearst, as a precedent for legislation stigmatizing and prohibiting marihuana on the federal level. Through this studied depiction of marihuana as a violent narcotic, the Federal Bureau of Narcotics effectively lobbied for the passage of the Marihuana Tax Act of 1937, which restricted the usage, distribution, and production of marihuana, thereby automatically administratively mandating the prohibition of the cultivation of cannabis hemp.

As recounted in a 1995 speech to the California Judges Association annual conference by Charles Whitebread, Professor of Law at USC Law School, entitled *The History of the Non-Medical Use of Drugs in the United States* (Bonnie & Whitebread, 1974), in the 1937 Marihuana Tax Hearing Dr. Munch stated in open court that he had conducted a series of experiments with cannabis on dogs and had, for scientific purposes, tried marihuana himself. When asked how it affected him, Dr. Munch testified, under oath, that, after two inhalations of a marihuana cigarette, he was turned into a bat, flew around the room, and down a 200-foot-deep inkwell.

Congress had never heard of marihuana and was unaware of its relationship with its hemp-cultivating constituents. But the hour was late, and it was time to move on. In a vote they didn't bother to record (Holfield, 2013), on a matter of little interest, to protect the nation from the threat of this violent narcotic, a handful of Congressmen ratified a bill that would one day fill the nation's prisons to the brims. During his act of demonization, Anslinger continuously cited his Gore file, the accumulated body of misinformation published by Hearst, as a precedent for legislation stigmatizing and prohibiting marihuana on the federal level. Through this studied depiction of marihuana as a violent narcotic, Harry Anslinger's

Federal Bureau of Narcotics effectively lobbied for the passage of the Marihuana Tax Act of 1937, which restricted the usage, distribution, and production of marihuana, thereby automatically administratively mandating the prohibition of the cultivation of cannabis hemp.

Cannabis hemp experienced a slight revitalization in the 1940s with the onset of the Second World War. Cut off from hemp fiber imports from the Philippines and Asia after Japan bombed Pearl Harbor, and upon realizing the war could not be won without the participation of *Cannabis sativa* (USDA, 1942), President Roosevelt signed an executive order that mandated emergency cannabis hemp cultivation to make essential supplies for the war effort, as these provisions were obligatory to provide cordage for the U.S. naval fleet. Hall (2022) recounted the U.S. Department of Agriculture instructing American farmers to ignore the federal law prohibiting the cultivation and harvesting of hemp. To portray what occurred more accurately, the United States did an end-run around the Marihuana Tax Act, which was not voided or amended. As a wartime emergency effort, by Presidential decree, The United States ignored Federal Law and demanded farmers to grow as much hemp as possible.

Hall (2022) revealed an unforeseen problem and an institutionalized solution related to this mandate. Many farmers had never grown hemp and had little understanding of how to produce the bumper crop the government was mandating. The solution was a short documentary film produced by the United States Department of Agriculture titled *Hemp for Victory!* (USDA, 1943). Hemp for Victory was described by Hall (2022) as a crash course in hemp cultivation. It was released in theaters in rural markets and distributed for non-theatrical screenings by agricultural agents working with farmers. Johnson (2014) reported that in 1943, hemp production in the United States reached more than 150 million pounds (140.7 million pounds of hemp fiber; 10.7 million pounds of hemp seed) on 146,200 harvested acres. Herer (1963) vividly described these supplies saving the life of a pilot who would eventually become an upper-echelon influencer and instrumental in the suppression of information about both the medicinal properties and industrial potential of cannabis hemp. When pilot George Herbert Walker Bush was forced to bail out of his burning airplane during a battle over the Pacific, he was unaware his aircraft engine was lubricated with cannabis hemp-seed oil, and the entirety of the webbing of the parachute that saved his life was made from cannabis cultivated in the United States. The rigging and the ropes and the fire hoses on the ship that pulled him from the ocean were woven from cannabis hemp, and as he stood safely on the deck, the stitching of his shoes was of cannabis hemp, as it is in all military shoes to this day.

World War II culminated in widespread fear of Communism and the Cold War (Patton, 2020), and hemp cultivation was no longer considered patriotic but was once more considered a threat, and the United States again became the only country in the industrialized world that outlawed the cultivation of cannabis hemp. In 1951, the Chief of the Los Angeles Police Department claimed communist agents were importing cannabis to corrupt American youth, and Anslinger's rhetoric evolved to conform with the McCarthy-era rhetoric that proliferated during this time to promote the anti-cannabis/hemp agenda. He informed the frightened American public that cannabis was much more dangerous than he initially thought. As the official expert on marihuana, he testified before a strongly anti-communist Congress in 1948, proclaiming that the substance did not, in fact, cause violence in its users but instead caused them to become peaceful and pacifistic (Herer, 1963) and that Communists could and would use marihuana to weaken our American fighting boy's will to fight. To combat the red menace, Congress voted to continue the prohibition of cannabis hemp based on the exact opposite reasoning it had used to outlaw it eleven years before.

Glaser (1956) recounted that in this era of McCarthyism, the Criminality Theory fell into disrepute among scholars and academics to be circumvented by the Gateway Drug Theory. Rather than botanic cannabinoid use being linked to addiction, insanity, and violent crime, the intromission of botanic cannabinoids was theorized to be a gateway to addiction to dangerous drugs like cocaine, heroin, and opioids. Based on the Gateway Theory, Congress ratified the Boggs Act, stipulating draconian mandatory prison sentences for violations of the Narcotic Drug Import and Export Act and the Marihuana Tax Act in 1951.

By the 1960s, the psychosocial structural medicinal botanic cannabinoid stigmatization conditioning of the American public by the upper echelon elite beginning in the wake of the Civil War had trickled down organically through society, causing the American public to be on guard against marijuana addicts and dealers, which in Anslinger's words were "nothing less than an assault on the foundation of Western civilization" and that "for this specific reason, the only persons who frighten me are the hippies" (McWilliams, 1989, p. 234).

Patton (2020) purported that throughout the 1960s and early 70s, American mainstream society had become suitably institutionally conditioned to categorize botanic cannabinoids as narcotics used by hippies, individuals that society viewed as occupying the lowest socioeconomic levels of society. The great social movements of the 1960s and early 1970s attributed to the hippies were akin to the crusades against other structurally stigmatized groups, including women, Native Americans, Hispanic Americans, homosexuals, the elderly, and various ethnic and minority groups.

Harry Anslinger resigned as commissioner of the Federal Bureau of Narcotics in 1962, and Musto and Korsmeyer (2002) described his later astonishment at the explosion of botanic cannabinoid use throughout the hippie era. By the end of the 1960s and into the 1970s, the structural stigmatization of botanic cannabinoid users moved higher into the upper-echelon administrative levels. Facing backlash early in his first term as President for his continuation of the Vietnam War, Richard Nixon had a particular vendetta against structurally stigmatized groups, particularly people of color and hippies.

The stigmatization of medicinal botanic cannabinoid use culminated in the early 1970s with the Nixon mandate that only the medicinal potential of synthetic cannabinoids was allowed to occur within the confines of the United States (Patton, 2020). Thus, a scientific paradigm psychosocially emerged over the course of a century, which viewed synthetic bullets as medicinal, and the naturally derived bullets with which humanity engaged in its war against disease and mood disorders for 12,500 years became stigmatized and were institutionally eliminated from humanity's medicinal arsenal.

However, an accumulation of anomalies, such as deaths and other adverse events resulting from synthetic cannabinoids, has resulted in challenges to this dominant paradigm (Tait et al., 2016). Thomas Kuhn

(1962) proffered the notion that the scientist's role is to design studies that could produce results that challenge the dominant paradigm and coined the word "revolution" to describe dramatic changes in scientific worldviews. Still, revolutionary science is torturous and painful because it shakes all confidence that science has in its present theories and underlying assumptions. Paradigm shifts occur gradually during a stage when the dominant paradigm is termed to be in "crisis." A new paradigm is currently emerging, which professes that botanic cannabinoids have medicinal properties without the side effects so prevalent through the ingestion of synthetic medicines (Alves et al., 2020).

This new paradigm is so simple that it has been taught in high school biology classes for decades. Whenever a molecule is ingested, whether that molecule is natural, like food, a botanic cannabinoid or terpene, or a synthetic molecule like an FDA-approved pharmaceutical medication, the body sends out an enzyme to degrade that molecule, break that molecule down, or euphemistically to eat that molecule. If that molecule is natural, like food or a botanic cannabinoid, that enzyme has undergone eons and eons of evolution to be adapted to eating that molecule, breaking that molecule down, or essentially degrading it. This is metabolism in action. However, if that molecule is synthetic, it is like that enzyme is trying to eat plastic fruit. So, the enzyme summons a worker molecule known as a protein and tells that protein that it cannot eat the molecule and to haul it away and store it somewhere. Usually, the protein hauls it to the liver, but sometimes it stores it in the kidneys, brain, or some other part of the body. Wherever it is stored, that synthetic molecule causes inflammation. Typically, when people think about inflammation, they think of pain. However, depending on the proinflammatory cytokines activated, inflammation may also manifest as mood disorders such as anxiety, depression, sleeplessness, and irritability, which, effectively, are the symptoms of PTSD (Sun et al., 2021).

The pursuit of scientific truth has historically led humanity to places unimagined. However, throughout human history, institutional constraints have been imposed on the areas of inquiry researchers are permitted to pursue, as well as their ability to share newly attained knowledge. For example, Socrates was executed for advocating a paradigm contrary to the institution to which he was mandated to adhere. Since Socratic times, scientists have been conditioned to respect and understand the dangers of pursuing research, which has the potential to challenge an institutionally established paradigm. Throughout the history of science, the persecution of scientists who subvert paradigms by institutions has been ubiquitous. As a science, physics has undergone multiple paradigm shifts throughout human history and the institutionalized persecution of the scientists responsible for subverting the dominant paradigms is ingrained in the historical record. For example, the Copernican Revolution entailed the shift from the geocentric (earth-centered) model of the solar system to the heliocentric (sun-centered) model in the 16th century. The Church, as the governing institution, condemned this paradigm because it demanded that biblical cosmology be taken literally. It considered the dissemination of this new paradigm heretical because it was contrary to the institution's interpretation of holy scriptures. Copernicus avoided persecution for this challenge to doctrine because he had the good fortune of dying, thereby avoiding being subjected to the torture implements utilized by institutionalized inquisitional functionaries.

In America, Richard Nixon established the precedent that politicians determine which aspects of truth scientists are permitted to pursue by prohibiting studies of naturally produced cannabinoids and only allowing research on cannabinoids that are produced synthetically. This prohibition on pursuing truth related to biologically produced cannabinoids lasted half a century. Ortiz and Preuss (2022) purported that Nixon declared the War on Drugs in general and biologically produced cannabinoids in particular for one specific reason: to decimate his perceived political enemies: people of color and hippies.

The first salvo by the Nixon Administration in the war against users of botanic cannabinoids occurred in 1968 when the now anemic Anslinger-less Federal Bureau of Narcotics was merged with the Bureau of Drug Abuse Control to form the Bureau of Narcotics and Dangerous Drugs, yet another division of the United States Department of Justice. Ortiz and Preuss (2022) described a proverbial nuclear weapon being deployed in 1970 with the ratification of the Controlled Substance Act (CSA), establishing a policy to regulate the use, manufacturing, distributing, importing, and exporting of federally regulated substances. The Act categorized these controlled substances and created a legally enforceable foundation for their regulation. In a historical review of the Federal Bureau of Narcotics and Dangerous Drugs, McWilliams (1989) described the next proverbial nuclear device being detonated in 1973 when this Federal Bureau was subsumed into the newly established Drug Enforcement Administration, a federal paramilitary organization tasked with establishing and enforcing controlled-substances laws of the United States.

The newly established United States Drug Enforcement Administration categorized chemicals, drugs, and various compounds used to make drugs into five classes or schedules depending upon their abuse or addiction potential and suitability for medicinal purposes. Kenny and Zito (2022) reviewed the development and implementation of the Federal Comprehensive Drug Abuse Prevention and Control Act and the claim by the DEA that the determining factor for placement of a substance on the Schedule is the potential for abuse. Schedule I drugs are alleged to have the highest potential for abuse and are purported to cause severe psychological and physical dependence. The lower the substance appears on the Schedule, the less potential the drug has to be abused. Robbins (2018) speculated that legal drugs such as alcohol and tobacco do not appear on the Schedule because they are outside the purview of the DEA, the paramilitary division tasked with adjusting the Schedule and enforcing it upon the citizenry. With binding input from the Secretary of the Health, Education, & Welfare Department, Bonnie (1974) purported that the Controlled Substances Act empowered the Attorney General to schedule new drugs and determine whether to re-schedule existing ones.

Disgraced Attorney General John Mitchell defied the recommendation of the National Committee on Marihuana and Drug Abuse and classified botanic cannabinoids as Schedule 1, categorizing them as the most dangerous molecules a human can ingest, possessing the highest degree of danger and illegality on the drug scale the Nixon administration created in his War on Drugs. In 1994, John Ehrlichman, Nixon's former domestic affairs advisor, and Watergate co-conspirator, described the strategy the Nixon administration used for engaging the enemy.

> We knew we couldn't make it illegal to be either against the war or Black, but by getting the public to associate the hippies with marijuana and Blacks with heroin and then criminalizing both heavily, we could disrupt those communities. We could arrest

their leaders, raid their homes, break up their meetings, and vilify them night after night on the evening news. Did we know we were lying about the drugs? Of course, we did. (Exum, 2019, p. 5).

The Schedule 1 classification had significant ramifications on academia and other institutional entities. To comply with the Controlled Substances Act, research universities were forbidden to engage in research on biologically produced cannabinoids, as conducting such research would jeopardize their federal funding (Mead, 1991). At the time, the only known biologically produced cannabinoids were the phytocannabinoids, the cannabinoids derived from the cannabis plant. This classification for the biologically produced cannabinoids worked well for the next two decades because it fit perfectly into the moral model of addiction, which fueled the Drug War (Waters, 2019). Dawson and Persad (2022) purported that during this time, the DEA flourished through asset forfeitures and massive budgeting allocations provided to ensure the Judicial System punished Americans for possessing cannabinoids produced biologically. The callousness of initiating the longest war in which America ever engaged as a strategy for abating the political damage done to him by the Vietnam War, Richard Nixon, an administrator influencing societal attitudes from the pinnacle of power, provided a strategy for the advancement of many political campaigns that succeeded his. Hawdon (2001) described President George H. W. Bush extending the war by equivocating botanic cannabinoids with heroin and crack cocaine in his first televised national address and referring to them as "the greatest domestic threat facing our nation today." It was during his presidential term that the United States National Institute of Health funded the development and utilization of the technology and instrumentation that led to the discovery of the biological cannabinoids the human body produces naturally, as well as the mechanism that causes botanic cannabinoids to demonstrate efficacy in the treatment of virtually every existing autoimmune disorder (Katchan et al., 2016).

Levitt et al. (1981), Sallan et al. (1980), and Ungerleider et al. (1982) detailed the accumulation of studies in the early 1980s that demonstrated botanic cannabinoids possessed significant medicinal properties. Still, the precedent that politicians make medical decisions for the citizens they govern had been established by Harry Anslinger. Faubion (2013, p. 405) recollected President Ronald Reagan exhibiting his medical expertise while perpetuating the structural stigmatization conditioning of America by stating, "I now have absolute proof that smoking even one marijuana cigarette is equal in brain damage to being on Bikini Island during an H-bomb blast."

It is currently difficult to accurately determine the number of Americans who are using botanic cannabinoids medicinally because research indicates that a significant percentage of the population is doing so covertly. Still, a substantial proportion of the North American population self-reported the use of botanic cannabinoids for medicinal purposes for various medical reasons, including those residing in states where their possession and intromission are prohibited. An International Cannabis Policy Study conducted by Leung et al. (2022) revealed that the overall prevalence of self-reported botanic cannabinoid use for medical purposes in the United States was 27%, with similar rates by sex and the highest prevalence in young adults. Prevalence was higher in US legal adult-use states (34%) than in US illegal states (23%). At the last census, the population of the United States was 333.29 million. Extrapolating from the findings of this study, simple math reveals that 89 million Americans are currently using botanic cannabinoids for their perceived medicinal properties. The most common physical health reasons for the medicinal use of botanic cannabinoids include pain management (53%), sleep (46%), headaches/migraines (35%), appetite (22%), and nausea/vomiting (21%). For mental health purposes, the most common uses were for anxiety (52%), depression (40%), and PTSD/trauma (17%). There were 11% who reported using cannabis for managing other drug or alcohol use and 4% for psychosis.

These numbers are arguably significant enough to indicate a diminished stigmatization of the medicinal use of botanic cannabinoids. Still, the combined theories of Goffman (1963) and Link and Phelan (2014) reveal that historically, once a stigma becomes psychosocially ingrained in American society, the stigma is never truly vanquished. Currently, hospitals and rehab centers provide areas where patients are allowed to combust and inhale dangerous and deadly molecules such as nicotine, but these healthcare facilities are institutionally mandated to prohibit patients from ingesting the medicinal molecules derived from the species *Cannabis sativa* because these cannot attain the glorifying label of FDA

Approved. While this violates every principle of biomedical ethics to which all healthcare institutions must adhere, this stigmatization by the American Healthcare Delivery System endures.

Because the research of biological cannabinoids was prohibited in the United States by the Controlled Substances Act, the National Institute of Health (NIH) provided funding for researchers at Hebrew University in Jerusalem who utilized newly invented technology and instrumentation to discover the endocannabinoid system (Parsons & Hurd, 2015), a biological system composed of endogenous lipid-based retrograde neurotransmitters (endocannabinoids) that bind to cannabinoid receptors and cannabinoid receptor proteins that are expressed throughout the vertebrate central nervous system and peripheral nervous system. Tate (2014) revealed that the discovery of this biological system in 1992 resulted in a rudimentary understanding of the medicinal properties of botanic cannabinoids and the psychological and biological efficacy of phytocannabinoid supplementation for endocannabinoid deficiency disorders and California voters enacting a legal and regulatory framework known as Proposition 215 in 1996, which allowed patients to possess and cultivate botanic cannabinoids for their personal medicinal use upon the recommendation of a medical doctor. Throughout the next decade, other states began allowing ailing and disabled individuals access to all federally illegal botanic cannabinoids (Johnson & Colby, 2023).

Mead (2019) provided an in-depth discussion of the convoluted and contorted interpretations of the legal and regulatory frameworks that surround botanic cannabinoids in the United States, including federal law as dictated by the Controlled Substances Act and governed by various federal agencies like the FDA and DEA, as well as state law, as regulated by each state's rules and regulations authorizing the use of botanic cannabinoids. Under the Controlled Substances Act, marijuana is defined as:

> The term "marihuana" means all parts of the plant Cannabis sativa L., whether growing or not; the seeds thereof; the resin extracted from any part of such plant; and every compound, manufacture, salt, derivative, mixture, or preparation of such plant, its seeds or resin. Such term does not include the mature stalks of such plant, fiber produced from such stalks, oil or cake made from the seeds

of such plant, any other compound, manufacture, salt, derivative, mixture, or preparation of such mature stalks (except the resin extracted therefrom), fiber, oil, or cake, or the sterilized seed of such plant which is incapable of germination. 21 USC 802(16) (emphasis added)

This definition reveals that marijuana includes its compounds and derivatives, as well as synthetic versions thereof. Therefore, each of the more than 100 botanic cannabinoids delineated by Karch (2007) is classified as Schedule I by operation of the definition and not because of any scientific analysis of their abuse potential. Only THC is separately and specifically listed in the Controlled Substances Act as a Schedule I substance.

Mead (2019) further delineated the convoluted and conflicting federal and state laws relevant to botanic cannabinoids and the subsequent confusion among every institution of which the American Healthcare Delivery System is comprised. Her analysis supports the research of Shi and Singh (2022), who described the American Healthcare Delivery System as an irrational and unintegrated network of components designed to work together incoherently. The interaction of these components has led to uncertainty and disarray among the FDA, healthcare providers, patients, diagnostic centers, nursing homes, medical equipment vendors, and community healthcare centers. Hurd (2020) described every component of the American Healthcare Delivery System as being completely unprepared for the massive explosion of the therapeutic use of cannabidiol. He depicted the CBD explosion as heavily focused on mental health disorders due to its broad actions on multiple systems, including modulation of opioid, serotonin, and adenosine transmission. The modulatory actions of CBD on these diverse biological systems undoubtedly affect multiple biopsychological functions, including anxiety, cognition, inflammation, mood, and nociception. The ubiquitous claims of CBD catalyzing homeostatic well-being stem from studies demonstrating the anxiolytic properties of this botanic cannabinoid.

Moreover, since it does not activate dopamine neurotransmitters, CBD is non-addictive. Taylor et al. (2019) reported consistent findings in clinical studies of cannabidiol, demonstrating that it is safe, well tolerated, lacks toxicity, and is nonintoxicating with no reinforcing properties. Even doses of up to 6000 mg ingested by healthy individuals resulted in no severe effects. The most notable adverse events were gastrointestinal, including diarrhea, with none of these side effects considered to be severe.

Alice Mead (2019), the Director of U.S. Professional Relations for GW Pharmaceuticals, a significant player in the healthcare delivery arena, recently described the company's Occam's razor approach in the pharmaceutical company's creation of the first and only FDA-approved medicine with a botanic cannabinoid-based component available in the United States. Its active ingredient is cannabidiol and is used to treat seizures associated with Lennox-Gastaut syndrome and Dravet syndrome in patients two years or older. Tyler (2000) revealed that by administrative policy, the FDA cannot approve proposed medicines that do not have a synthetic (patentable) component. GW Pharmaceuticals exploited this policy by attaching a synthetic strawberry-flavored molecule to a cannabidiol phytocannabinoid and patenting the preparation as Epidiolex. The compound was submitted to the FDA as an orphan drug to treat Lennox-Gastaut syndrome, a rare, severe form of epileptic encephalopathy, frequently treatment-resistant to conventional pharmaceutical anti-seizure medications. Thiele et al. (2018) depicted the drug as comprised of 98% botanically derived cannabidiol, with 2% of the compound consisting of a laboratory-created strawberry-flavored molecule called Sucralose. Sucralose served two purposes. First, it made the medicine palatable for children. More importantly, this synthetic component allowed the concoction to be eligible for FDA approval.

Adverse psychological effects of Epidiolex were absent in the clinical trials the British pharmaceutical company presented to the FDA. The physiological side effects of Epidiolex were listed by Thiele et al. (2018 pp. 1090-1091) as "diarrhea, vomiting, acute hepatic failure, hepatic failure, viral infection, increased concentration of another antiepileptic drug in the blood, convulsion, lethargy, restlessness, acute respiratory distress syndrome, acute respiratory distress, acute respiratory failure, hypercapnia, hypoxia, pneumonia, aspiration, and rash," Moore and Furberg (2014) purported that by precedent, the FDA considers these iatrogenic effects insignificant and consistently approves medicines with side effects such as these, which they categorize as minor. The synthetic component combined with the precedent of approving medications with similar side effects implies the federal agency had no choice but to pronounce Epodiolix 'safe and effective' (Howland, 2008). Still, the differences in side effects between CBD alone and CBD combined with sucralose suggest that the adverse effects come from the synthetic component.

Summary

Botanic cannabinoids were human civilization's first effective widespread medicines, having been used medicinally for over 12,500 years. It was not until the advent of synthesized medicines that humans began stigmatizing these molecules and the people who use them. In 1922, Congress ratified the Narcotic Drug Import and Export Act and created the Federal Narcotics Control Board, defining "narcotic" as "any drug used by individuals of low socioeconomic standing" (Patton, 2020, p. 7). This Act was the first recorded instance of the United States stigmatizing a substance based on the user's socioeconomic status, and the stigmatization of botanic cannabinoid use continued to manifest in innumerable forms from multiple entities throughout the next century.

Chapter 4

Cannabis Chemovars and a Holistic Perspective of Why Cannabis Works as Medicine

Scientists like to classify things, and cannabis is no different. Due to difficulties in distinguishing one cannabis cultivar from another based-on factors such as plant height and leaflet width and the fact that all cannabis types are eminently capable of crossbreeding to produce fertile progeny, the only reasonable solution is to classify them by their biochemical and pharmacological characteristics while referring to stains of cannabis as "chemovars" or chemical varieties (Russo, 2019). These chemovars are analogous to phytochemical factories producing terpenes and cannabinoids, many with well-documented medicinal properties (Baron, 2018).

When it comes to classifying medicinal cannabis, there is a popularized method that results from politicians making medical decisions. As the medicinal use of the molecules produced by the genus *Cannabis* became politically acceptable, institutions permitted to cultivate and distribute *Cannabis* became mandated to classify it based on the quantity of intoxicating cannabinoids produced. This resulted in regulated and institutionally acceptable terminologies for classifying the genus Cannabis, such as indica, sativa, hybrid, hemp, and drug type that are regulatory mandated by the state institutions that allow Cannabis cultivation and distribution at the state level (Lapierre et al., 2023).

Regulators in states that allow Cannabis cultivation and distribution have conditioned individuals to consider the plant according to the amount of the intoxicating cannabinoid delta-9-tetrahydrocannabinol (THC) the cultivar produces. Plants that produce less than 0.3% THC are institutionally classified as hemp, while plants that produce 0.3% or more THC are categorized as a *drug type* (Monthony et al., 2021; Torkamaneh & Jones, 2022). THC has been the main point of focus for the institutions that craft the legislation that regulates botanic cannabinoids, despite the fact that the genus *Cannabis* produces over 120 different cannabinoids (Mudge et al., 2018; Hurgobin et al., 2021).

The Indica/Sativa Classification Method

The Indica/Sativa classification method is the system that all medical cannabis patients should be familiar with because it is used by American dispensaries and has been used for decades. This is the method that all medical cannabis patients should be familiar with because it has been used for decades and is so simple that everyone understands it. It is also extremely useful when beginning to learn about medicinal cannabis. Essentially, this classification system groups cannabis into three categories: Indicas, Sativas, and hybrids of the two.

Sativas are typically described as uplifting and energetic, whereas Indicas are described as relaxing and calming. However, the distinction between Sativa and Indica, as commonly applied in popular literature, is total nonsense and has been described as an exercise in futility (Piomelli & Russo, 2016). The degree of interbreeding hybridization of cannabis chemovars is such that only a biochemical assay provides a potential consumer or scientist an accurate assessment of the medicinal molecules the plant contains. The differences in observed effects in cannabis chemovars are due to their cannabinoid and terpene content. Terpenes are rarely assayed, let alone reported to patients. The sedation effect of the so-called Indica strains is erroneously attributed to CBD content when, in reality, CBD is stimulating in low and moderate doses. Instead, sedation in most common cannabis chemovars is attributable to the myrcene content, a monoterpene with a strongly sedative couch-lock effect that resembles a narcotic. Biochemically, any cannabis plant containing more than 0.005% myrcene is classified as an Indica (Hazekamp & Fischedick, 2012). Both Indicas and Sativas are "narcotic-like" in that they control pain by inhibiting pain receptors in the body from firing pain messages to the brain.

Both Indicas and Sativas are "narcotic-like" in that they control pain in much the same way as narcotics do (inhibiting pain receptors in your body from firing pain messages to the brain). What cultivators of cannabis have done is bred these two strains of cannabis in a variety of ways to produce hybrids. Sativa and Indica hybrids are purported to produce slightly different physiological and psychological effects, as reflected in this table.

Hybrids Explained

Sativa-Dominant Hybrids	Indica-Dominant Hybrids
• Activating strains for daytime use.	• Sedating strains for use at night.
• Provide a feeling of well-being.	• Provides full-body pain relief and relaxes muscles.
• Cerebral high with an energizing body effect.	• Perfect for patients who suffer from autoimmune disorders, insomnia, or depression.
• These strains provide both physical and mental relief.	• Reduces nausea and relieves migraines and headaches.
• Tend to be uplifting with cerebrally focused effects.	• Relieves anxiety and stress.
• Promotes creativity and boosts imagination.	• Helps to reduce pain and inflammation.
• Good for pain, fatigue, and mood disorders.	

The way most medical cannabis patients medicate is this: in the morning they medicate with an activating strain of sativa. Selecting an appropriate sativa dominant-hybrid allows them to be productive and to do the variety of things that a person needs to get done during the day. These strains tend to promote creativity and boost imagination. This is the reason why many artists, philosophers, and musicians use sativa dominant strains.

Indica strains tend to be used at night. They are relaxing and are useful when you just want to kick back and watch a movie or your favorite Netflix marathon or just simply want to go to sleep.

The Indica/Sativa classification system is one that everyone uses and understands, and so is the one every medical cannabis patient should internalize. The problem is that genetic analysis has proven that while this classification system may be useful, it is not actually true.

Since the 1970s, cannabis has been divided into three sub-species (often confused as different species), *Cannabis Indica, Cannabis Sativa,* and *Cannabis Ruderalis,* with ruderalis principally being considered 'wild cannabis' and not suitable for medicinal or recreational purposes. An institutionalized distinction was created by politicians between cannabis, which is bred for high THC content, and hemp, which is typically bred for industrial uses like fiber, rope, and paper.

Richard Evans Schultes et al. (1974) created the original taxonomy for cannabis in the 1970s, misidentified a *Cannabis afghanica* plant as a *Cannabis indica* plant. This mistake created 40 years of confusion, which was dispelled by the research of McPartland and Guy (2017). McPartland and Guy were the first researchers to examine the genetic markers on the three subspecies of cannabis using the plant's genome to conclusively identify where it originated. They also proved convincingly that these are all the same species, just different subspecies. As it turns out, *Cannabis sativa* should have been identified as *Cannabis indica* because it originated in India (hence *indica*). *Cannabis indica* should have been identified as *Cannabis afghanica* because it actually originated in Afghanistan. Finally, it seems that *Cannabis ruderalis* is actually what people mean when they refer to *C. sativa.* If this seems confusing, McPartland and Guy (2017) simplified things by providing a chart:

Cannabis Indica (Formerly Sativa)

Origin: India

Morphology: Taller (>1.5m) than their short and stocky Afghanica cousins, with sparser branches and less dense buds/flowers.

Physiology: Longer flowering time, between nine and fourteen weeks. Minimal frost tolerance with moderate production of resin.

Chemistry: Much greater THC than CBD and other cannabinoids, this leads to the "head high" many users report.

Psychoactivity: Stimulating.

Cannabis Afghanica (Formerly Indica)

Origin: Central Asia (Afghanistan, Turkestan, Pakistan)

Morphology: Shorter (<1.5m) than Indica strains with dense branches with wider leaves, and much denser buds/flowers

Physiology: Shorter flowering time, as little as seven to nine weeks. Good frost tolerance with high resin production. Afghanica strains can be susceptible to mold due to how dense the buds and branches are.

Chemistry: More variable than Indica strains. THC is often still the predominant cannabinoid but some strains have 1:1 ratios and some may have even higher CBD than THC.

Psychoactivity: Sedating.

Cannabis Sativa (Formerly Ruderalis)

Origin: Usually feral or wild. From Europe or Central Asia.

Morphology: Variable, depending on origin.

Physiology: The flowering time is short and variable. Many varieties exhibit auto-flowering traits (flowering independently of sun cycles). Moderate frost tolerance with relatively low resin production.

Chemistry: More CBD than THC. Prominent terpenes include caryophyllene and myrcene, giving these strains a floral flavor and scent.

Psychoactivity: Usually lacking.

This new nomenclature should arguably replace the old system because it is grounded in the actual genetics of the plant and is scientifically sound. Despite this, it is likely that this new classification system will face resistance from cannabis users and those in the medical cannabis industry who will have become used to decades of convention firmly establishing an inaccurate taxonomy. This is reminiscent of the Brontosaurus, a dinosaur that never existed but we were all taught in school to believe was real. Sometimes, science gets it wrong, and it is the responsibility of modern scientists with better methods, like McPartland, to correct the old mistakes.

The difficult part will be getting mass acceptance of his newly proposed taxonomy. It seems likely that a split will develop between academics and marketers of the plant, with academics adopting the new system and dealers continuing to adhere to the old system for many years to come. Perhaps in time, *Cannabis afghanica, Cannabis indica, and Cannabis sativa* will come into fashion, but that basically will depend on the willingness of the medical cannabis industry to adopt this new system and thus pass it on to the patients and growers. But it seems unlikely that the cannabis industry will wholeheartedly jump on board, given the risk that this new nomenclature could confuse patients who may be used to seeing only "Indicas" and "Sativas" on the shelves. So, while the indica/sativa classification system has been used for decades and likely will be used for decades more and is the one every medical cannabis patient should learn, be aware that it has no real basis in reality.

Chapter 5

The Relationship of the Endocannabinoid System to Autoimmune Disorders and Addiction

The endocannabinoid system (ECS) is a widespread neuromodulatory network involved both in the functioning of the central nervous system and playing a significant function in mitigating many cognitive and physiological processes. The endocannabinoid system is comprised of endogenous cannabinoids (endocannabinoids), cannabinoid receptors, and the enzymes responsible for the production and degradation of endocannabinoids. The endocannabinoid system performs multiple functions in the vertebrate body, and the goal of this system is to maintain a stable internal environment. Put, the endocannabinoid system serves as an internal homeostatic regulator consisting of three main components: cannabinoid receptors, endocannabinoids, and the enzymatic machinery responsible for endocannabinoid synthesis and degradation (Piazza et al., 2017).

Endocannabinoids

Narrowly defined, endogenous cannabinoids (endocannabinoids) are signaling lipids that activate or deactivate cannabinoid receptors. 2-arachidonoyl glycerol (2-AG) and anandamide are the two best-known, but other structurally related lipids also engage cannabinoid receptors. These include Virodhamine, Oleoylethanolamide, and N-arachidonoyl dopamine (Devane et al., 1992; Mechoulam et al., 1995; Grabiec & Delghani, 2017; Lambert & Muccioli, 2007).

Cannabinoid Receptors

There is a widespread network of cannabinoid receptors throughout the body, including the best-characterized cannabinoid receptor types, CB1 and CB2. Both the CB1 and CB2 are G protein-coupled receptors (GPCRs) that predominantly bind to inhibitory G proteins. The range of CB1 signaling is enhanced by their propensity to heterodimerize with other GPCRs, including D2 dopamine, hypocretin, and opioid receptors (Lu & Mackie, 2021).

The CB1 receptors outnumber most of the other receptor types in the brain and central nervous system, while the CB2 receptors are expressed throughout the peripheral immune system. CB1 receptors are found in particularly high levels in the neocortex, hippocampus, basal ganglia, cerebellum, and brainstem (Herkenham et al., 1991; Marsicano & Lutz, 2006). With one notable exception, activation of CB1 receptors induces a decrease in the release of most major neurotransmitters in the central nervous system, including GABA, acetylcholine, noradrenaline, and serotonin (Castillo et al., 2012; Kano et al., 2009; Schlicker & Kathmann, 2001). The exception is that CB1 receptor agonists activate mesencephalic dopaminergic neurons in the ventral tegmental area, resulting in a release of dopamine in the nucleus accumbens and the prefrontal cortex (Cheer et al., 2004; Chen et al., 1990; Diana et al., 1998; Gessa et al., 1998; Maldonado et al., 2006; Melis & Pistis, 2012; Patel et al., 2003; Tanda et al., 1997; Wenzel & Cheer, 2014).

CB2 expression is highly inducible on the reactive microglia in the central nervous system following inflammation or injury. CB2 receptors are widely and strongly expressed in a range of leukocytes and appear to be the key mediator of cannabinoid regulation of inflammation and immune functions (McKallip et al., 2002). This demonstrates that CB2 receptors are critical to helping control immune functioning and that these receptors play a role in modulating intestinal inflammation, contraction, and pain in inflammatory bowel conditions (Bie et al., 2018; Wright et al., 2008). Activating the CB2 receptors mitigates central neuroinflammation while suppressing reactive microglia behavior, and CB2 receptor activation has the potential to provide the anti-inflammatory effects of cannabinoids sans the pleasurable feelings of CB1 activation that is federally prohibited for psychosocial rather than medicinal reasons.

The classic concept of the endocannabinoid system has been extended with the discovery of other receptors, such as the family of peroxisome proliferator-activated receptors (PPARs) and endocannabinoid-like mediators such as palmitoylethanolamide and oleoylethanolamide. These are classified as endocannabinoid-like molecules because they are synthesized and metabolized by the same class of enzymes as anandamide but only in part share the same mechanism of action (Mc Partland et al., 2014; Petrosino & Di Marzo, 2017).

Metabolic Enzymes Responsible for the Synthesis and Degradation of the Endocannabinoids

Both anandamide and 2-AG contain arachidonic acid. Still, their routes of synthesis and degradation are completely distinct and are mediated by different enzymes. Anandamide is produced from N-arachidonoyl phosphatidyl ethanol (NAPE), whereas 2-AG is

produced from 2-arachidonoyl-containing phospholipids, primarily arachidonoyl-containing phosphatidyl inositol bisphosphate (Figure 1). Synthesis of anandamide appears to occur by

multiple pathways that vary depending on brain regions, and different pathways may be favored for distinct physiologic and pathophysiologic processes (Pacher et al., 2006). The synthetic pathways for 2-AG are less complex than the pathways for anandamide, with most 2-AG endocannabinoids appearing to be created by the sequential hydrolysis of an arachidonoyl containing phosphatidyl inositol bis-phosphate (Jung et al., 2007; Shonesy et al., 2015).

The main enzyme for the degradation of anandamide in the central nervous system is the enzyme fatty-acid amide hydrolase (FAAH). FAAH degrades multiple fatty acid amides, including palmitoyl and oleoyl ethanolamide. This has important therapeutic implications, as inhibition of FAAH increases levels of these ethanolamides, which have widespread actions independent of the cannabinoid receptors (Luchicchi et al., 2010). Degradation of 2-AG is primarily due to three hydrolytic enzymes: monoacylglycerol lipase (MGL), alpha/beta domain containing hydrolase 6 (ABHD6), and alpha/beta domain containing hydrolase 12 (Blankma et al., 2007).

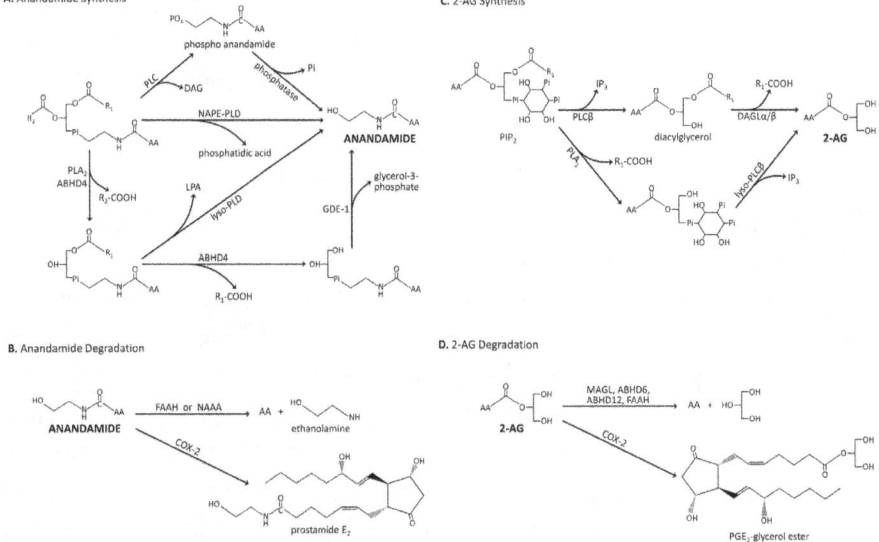

Figure 1. Potential synthetic and degradative pathways for anandamide and 2-arachidonoyl glycerol (2-AG). (A) Primary synthetic pathways for anandamide. (B) Primary degradative pathways for anandamide. (C) Primary synthetic pathways for 2-AG. (D) Primary degradative pathways for 2-AG.

The Endocannabinoid System and Its Relationship to Addiction

Experiments by Devane et al. (1992) and Parsons and Hurd (2015) revealed that the two most critical endocannabinoids related to addiction are anandamide and 2-AG. Anandamide is the endocannabinoid that agonizes the CB1 receptors responsible for activating dopamine neurotransmitters. This results in the sense of pleasure referred to as psychoactivity. The bliss resulting from dopamine activation has been demonstrated to be the biomolecular mechanism that makes psychoactive drugs addicting. 2AG is the endocannabinoid responsible for serotonin activation (Mechoulam et al., 1995), the neurotransmitter that provides a sense of happiness, contentment, and well-being (Lustig, 2017). For these reasons, from a biomolecular perspective, addiction treatment is more likely to be effective by activating serotonin while repressing dopamine. The most useful phytocannabinoid for accomplishing this task is cannabidiol. CBD inverse agonizes the CB1 receptor (Müller & Homberg, 2015), blocking its access and thereby mitigating the effects of addictive drugs. Simultaneously, it agonizes the CB2 receptors, which activate the serotonin neurotransmitters (De Gregorio et al., 2019). This biomolecular mechanism explains how CBD's intromission effectively treats every substance abuse disorder studied.

Systematic reviews of multiple studies by Prud'homme and Cata (2015) and Siklos-Whillans et al. (2021) indicated that biomolecular manipulation of the endocannabinoid system through phytocannabinoid supplementation would complement the Matrix Model, an integrative therapeutic approach incorporating the most efficacious aspects of behaviorism, person-centered therapies, cognitive behavioral therapy, the twelve-step approach, and motivational interviewing. As a harm reduction strategy, phytocannabinoid supplementation is supported by evidence demonstrating its efficacy for pain relief and as a substitution for multiple illicit drugs, alcohol, tobacco, and pharmaceuticals. Animal models demonstrate that phytocannabinoids reduce withdrawal, which contributes to drug-seeking behavior (Hurd et al., 2015), but statistically significant human clinical trials are lacking. With few exceptions, animal trials involved the addictive drug being administered intravenously, followed by the animals receiving a CBD systemic injection. While this methodology is deemed acceptable for rats and mice because they are excluded from the Animal Welfare Act, such an approach might not pass IRB scrutiny for human clinical trials in many industrialized nations (Banerjee, 2015).

The Endocannabinoid System and Its Relationship to Autoimmune Disorders

The immune system protects against disease-causing organisms by releasing antibodies which are proteins in the blood designed to neutralize a threat. The immune system is the system the body uses to prevent or limit infection. Its network of cells, organs, tissues, and proteins empowers the immune system to defend the body against pathogens. Occasionally, the immune system mistakes healthy cells for pathogens and sends antibodies to attack them. This is what happens with autoimmune disorders.

A fully functional immune system distinguishes healthy tissue from unwanted substances. If it detects an unwanted substance, it initiates an immune response, a complex attack to protect the body from bacterial, viral, and parasitic pathogens. It also recognizes and eliminates dead, damaged, and defective cells.

However, the immune system sometimes makes mistakes. Autoimmune disorders occur when the immune system becomes activated, mistakenly perceives the body's healthy tissues as invaders, and launches an unnecessary attack, leading to uncomfortable and sometimes dangerous symptoms. The human immune system is facilitated by beta cells, which manufacture antibodies to foreign invaders. Beta cells produce antibodies to prevent infection or rebuff bacterial, fungal, parasitic, and viral pathogens.

There are currently more than 80 recognized autoimmune disorders affecting various parts of the body, and more conditions are considered to be autoimmune-related. The reason the self-attack switch is activated is unknown, but the consensus among clinicians and physicians is that those with autoimmunity have a genetic predisposition for these conditions. Then, a specific event like an infection, parasite, leaky gut syndrome, or a traumatic experience triggers the autoimmune response.

Of the 23 million sufferers of autoimmune disorders in the United States, 75% are female, but it is unclear why. Females, in general, are thought to have a stronger immune response because men are about twice as likely to get cancer and infections. The stronger immune response of females is a mixed blessing. The enhanced immune system offers protection, but it predisposes females to an out-of-control immune system. This provides an explanation for why women are more susceptible to autoimmune conditions like fibromyalgia, rheumatoid arthritis, multiple sclerosis, and scleroderma.

Conventional pharmaceutical ideology believes that the immune system cannot be controlled and that once the autoimmunity switch has been triggered, it is impossible to revert back to the body's normal state. However, many physicians and clinicians believe that autoimmune conditions may be reversed or greatly resolved nutraceutically through the use of the medicinal terpenes and cannabinoids produced by the species *Cannabis sativa.* In typical treatment protocols, patients are not given information about these molecules because they cannot be patented and, therefore, cannot attain FDA approval. In general, patients experiencing autoimmune-related conditions are erroneously told by functionaries of the American Healthcare Delivery System that synthetic medications are required to get better.

Prescription medications synthesized to treat autoimmune disorders are designed to disable the immune system entirely. Immunosuppressive drugs are synthetically created antibodies that attack the autoimmune antibodies, and like all medicines created synthetically, they result in side effects because of the body's inability to degrade them. These synthetic medications may make the individual more susceptible to infection and frequently result in the development of cancer. Furthermore, immunosuppressive therapy is incredibly expensive and frequently ineffective, and many patients who try immunosuppressive medications discontinue them due to their iatrogenic effects. A new paradigm of medicine is emerging that purports that rather than obliterating the immune system, the aim should be to modulate the immune system and bring it back into balance rather than to shut it down completely.

Botanic cannabinoids and terpenes modulate the immune system, have proven efficacious for treating autoimmune disorders, and are safer, cheaper, and more effective in treating autoimmune conditions than pharmaceutical medications (Katchan et al., 2016). This accounts for the explosion in the number of medical cannabis patients in the United States since the discovery of the endocannabinoid system. The medicinal molecules produced by the cannabis plant modulate the immune system through their actions as Agonists, Partial Agonists, Antagonists, and Inverse Agonists.

Agonists, Partial Agonists, Antagonists, and Inverse Agonists

Agonists, partial agonists, antagonists, and inverse agonists are crucial, confusing, and convoluted terms used in the field of psychoneuroimmunology (Figure 2). *Agonists* (sometimes called full agonists) are molecules or chemical compounds that bind to and activate receptors, thereby producing a biological response (Gaynor & Muir, 2015). *Partial agonists* are molecules or chemical compounds that bind to and weakly activate a receptor, thus producing a submaximal biological response (Waller & Sampson, 2017). *Inverse agonists* are molecules or chemical compounds that bind to the same receptor site as an agonist and produce a biological response opposite to that of the agonist. Put more simply, inverse agonists bind to a receptor but deactivate it, causing a reduction in the baseline activity of the receptor (Bardal et al., 2011).

Antagonists are molecules or chemical compounds that bind to and block a receptor but do not activate it and thus produce no biological response. Antagonists occupy the receptor site and prevent the binding of agonists and inverse agonists, thus blocking their actions (Katzung & Trevor, 2020).

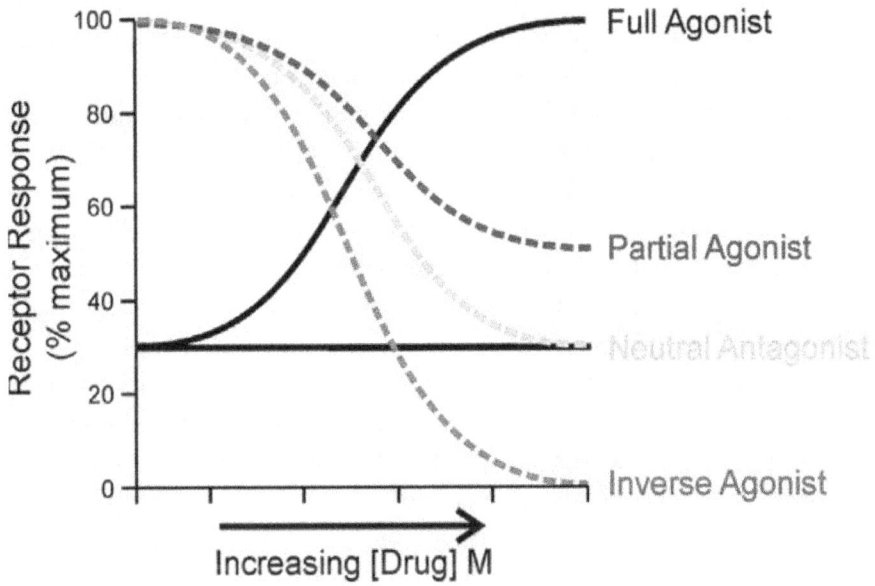

Effect of partial agonists, neutral antagonists, and inverse agonists on full agonist-stimulated receptor responses (Ferguson & Feldman, 2014)
License: CC BY-NC-ND 4.0

Cannabinoid receptor agonists are heterogeneous and are generally divided into four groups based on the differences in their molecular structure. These are classical, non-classical, aminoalkylindole, and eicosanoid compound groups (Howlett et al., 2002; Pertwee, 2005). The classical group consists of the botanic cannabinoids Δ9-Tetrahydrocannabinol (THC), cannabinol, and their synthetic equivalents. Synthetic equivalents to the botanic cannabinoids will not be delineated because, like any synthetic drug, they produce iatrogenic effects due to the body's inability to degrade them.

Autoimmune disorders are a group of ailments that the pharmaceutical industry fails to treat efficaciously and remain largely misunderstood. Cannabinoids (endocannabinoids and phytocannabinoids) have been shown to have a variety of effects on body systems. Through CB1 and CB2 receptors, botanic cannabinoids exert an effect by modulating neurotransmitter and cytokine release. Research on the role of botanic cannabinoids in the immune system demonstrates that they possess immunosuppressive properties. They can inhibit the proliferation of leucocytes, induce apoptosis of T cells and macrophages, and reduce secretion of pro-inflammatory cytokines. In mice models, they are effective in reducing inflammation in arthritis and multiple sclerosis and have a positive effect on neuropathic pain and type 1 diabetes mellitus. They are effective as a treatment for fibromyalgia and have been shown to have anti-fibrotic effects in scleroderma. Botanic cannabinoids can, therefore, be promising immunosuppressive and anti-fibrotic agents in the treatment of autoimmune disorders. Cannabinoids have been demonstrated to have immunomodulatory effects, and therefore, their potential role as an autoimmune and inflammatory treatment has been extensively investigated in scientifically progressive countries. Studies involving human models in the United States are scarce due to the Richard Nixon-initiated prohibition of research on cannabinoids not produced synthetically, and more research is required on cannabinoids produced biologically now that the Nixon Mandate has been rescinded.

Chapter 6

The Endocannabinoid System and its Relationship to Dosing with Cannabis

The endocannabinoid system was first characterized in the early 1990s and is a very complex bodily system that controls or influences all other bodily systems. The endocannabinoid system is still being mapped, but it is currently known to support such functions as mood, appetite, memory, digestion, bone cell growth, motor function, immune response, motor function, blood pressure, and the protection of neural tissues. The endocannabinoid system has been called the single most important scientific medical discovery since the recognition of sterile surgical techniques, and as our understanding of its complexities expands, it becomes increasingly evident that the endocannabinoid system controls all physiological and psychological processes (Hatzenbuehler, 2016).

Known as the endogenous cannabinoid system, this receptor network within our body is responsible for general health and is arguably our body's most important physiological system. It has a multitude of tasks and acts as a homeostatic regulator, making it directly responsible for maintaining homeostasis - the regulation of and maintenance of a stable environment within the human body. Currently, endocannabinoid system balance is considered to be related to the relative contribution of CB1 and CB2 receptors at any given time. Research is accumulating that demonstrates that CB1 dominance is associated with increased perception of stress, anxiety, paranoia, augmented appetite, decreased nausea and vomiting, and pain perception, as well as enhanced immune surveillance, the latter of which has been demonstrated in certain cancer models (Castillo et al., 2012; Pacher et al., 2006; Sharkey & Wiley, 2016). In contrast, CB2 dominance is associated with decreased inflammation and tissue injury in conjunction with improved metabolic health, insulin signal/sensitivity, satiety, and energy balance (Borgonetti et al., 2022; Patel & Hillard, 2009).

Recent biochemical and behavioral findings demonstrate that "optimal" activation of the CB1 receptors promotes anti-depressant-neurochemical changes and behavioral effects consistent with antidepressant/anti-stress behaviors in rodents (Patel & Hillard, 2009). These findings reinforce the importance of a balanced endocannabinoid system. Keep in mind that "balance" is critical, as research has shown that if we tip the scales too heavily in the direction of CB1 inhibition, there may be an associated decrease in fertility, with an increased risk of depression, mood disturbance, and immunosuppression. An overabundance of CB1 signaling has been associated with psychoactivity, systemic inflammation, cardiovascular risk, diabetes, and obesity (Chandy et al., 2024). In contrast, CB2 over-activation and dominance could lead to decreased immune function and diminished wound healing (Vuic et al., 2022).

The botanic cannabinoids produced by the cannabis plant are extremely similar to those that the vertebrate body naturally produces - the endocannabinoids. When botanic cannabinoids are introduced into the vertebrate body, interacting with the endocannabinoid system, the physiological, psychological, and medicinal effects of botanic cannabinoids are experienced. A basic way of conceptualizing this is that the body creates endocannabinoids to regulate and modulate its system. When we introduce botanic cannabinoids, we are effectually putting the system into overdrive by increasing its cannabinoid presence.

As has already been discussed, the endocannabinoid system comprises different receptors, each reacting with cannabinoids in different ways but with the same goal: homeostasis. It is the way these receptors bind and interact with the different botanic cannabinoids that cause the holistic and profound effects we can feel from these nutraceutical molecules. It is the way THC, CBD, and the minor botanic cannabinoids interact with our body chemistry, quite often to our benefit, that causes these natural molecules to have such comprehensive and encompassing implications in medical science. Anyone with clinical signs or symptoms of obese-central metabolic syndrome, pain, anxiety, depression, sleep disturbance, or inflammatory-immune dysfunction will likely demonstrate some endocannabinoid system imbalance that can be part of a multifaceted therapeutic strategy to manage these conditions in healthy (or not-so-healthy) individuals.

The botanic cannabinoids produce a balanced ecology of chemical compounds that play well with a host of systems within our body that are controlled by the endocannabinoid system. Botanic cannabinoids act as homeostatic regulators for the endocannabinoid system. This means that they act to keep the endocannabinoid system in balance, and proper dosing ensures that the metaphorical scales don't tip too far one way or the other.

Dosing with Cannabis

This balance of the endocannabinoid system is directly related to dosing with medical cannabis. When you ask people who think they know cannabis about dosing, you generally get bad advice. Typically, people's perception of cannabis dose is determined by how much they can withstand.

This is the stupidest, most pigheaded way of approaching a botanical

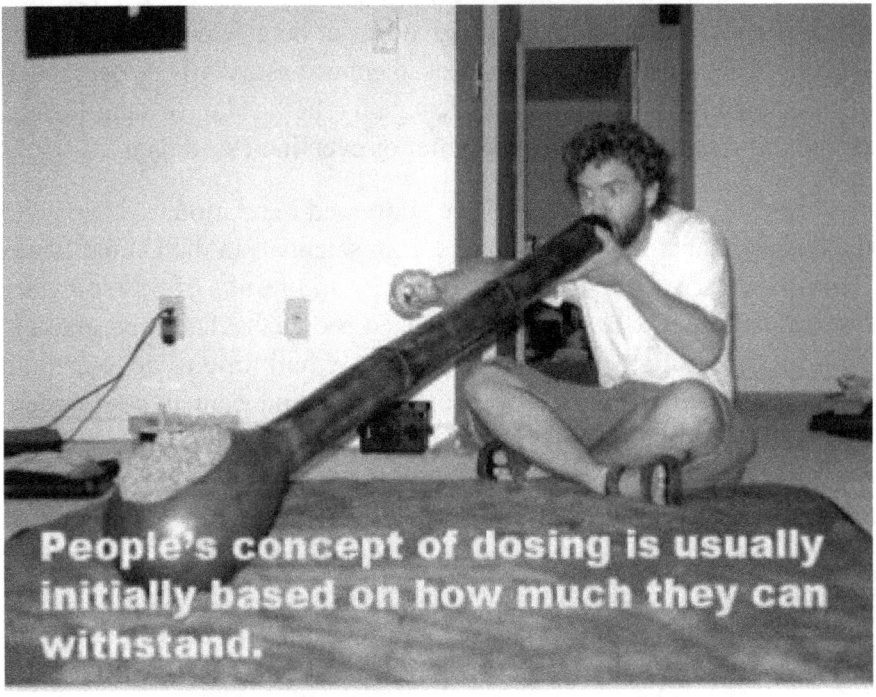

People's concept of dosing is usually initially based on how much they can withstand.

substance you can imagine. Nobody thinks you're a better person because you can drink a whole bottle of Tequila. Multiple studies have demonstrated that moderate doses of cannabis are particularly effective for most levels of neuropathic pain. Many cancer patients who are often at higher pain levels may require larger doses for pain control. The point is that dosing at too high a level can throw the endocannabinoid system out of balance and should be avoided.

Three studies are particularly relevant when it comes to dosing with medical cannabis. The first study was conducted by researchers in Israel (Wilsey et al., 2013), and the undeniable conclusion of the results was that low doses of cannabis work effectively well against most levels of neuropathic pain such as those experienced by sufferers of arthritis, muscular dystrophy, diabetes, fibromyalgia, and many other painful conditions. This conclusion is reinforced by another study conducted by researchers in Spain. Wallace et al. (2007) injected volunteers with capsaicin (which is the active ingredient in Bengay cream). The result of doing this provides a particularly good model for pain. They then administered low, medium, and high doses of botanic cannabinoids and terpenes. They found that at low levels, there was essentially no pain relief; at moderate levels, there was excellent pain relief; and most significantly, at high levels, there was no pain relief or even increased pain.

The final study that needs to be addressed in relation to dosing and the endocannabinoid system comes from scientists in the United States (Hirvonen et al., 2012). The methodology and results of this study are fascinating because the scientists collected people that had been smoking 30 joints a day for years and so obviously had time to devote to an in-patient study. These are also people whose endocannabinoid system was well downregulated (out of balance) because they were ingesting so much THC. To comply with U.S. policy related to research on non-synthetic cannabinoids, the researchers were attempting to prove that botanic cannabinoid use causes brain damage because the only way to get permission to study biologically produced cannabinoids in America is by attempting to prove that they cause harm. This study essentially entailed locking the test subjects up for 28 days and not allowing them access to botanic cannabinoids. The result of the study was that after the 28 days were up, their endocannabinoid system returned to normal baseline levels in all the participants, as documented by brain scans. This study demonstrates that you can smoke 30 joints a day for years, stop for 28 days, and your endocannabinoid system essentially resets itself.

An Analogy

A useful analogy to describe how botanic cannabinoids work in the human body is to consider the human body as a grand corporation of chemicals (carbohydrates, proteins, lipids, and enzymes). These represent the middle managers and laborers. These compounds all work together synergistically to maintain your company's metabolism and must be balanced to work together efficiently. The amalgamation of botanic cannabinoids plays the role of the consultant, coming in occasionally when things get a little out of whack to reestablish balance in the company and keep it headed in the right direction.

is that dosing at too high a level can throw the endocannabinoid system out of balance and should be avoided.

Three studies are particularly relevant when it comes to dosing with medical cannabis. The first study was conducted by researchers in Israel (Wilsey et al., 2013), and the undeniable conclusion of the results was that low doses of cannabis work effectively well against most levels of neuropathic pain such as those experienced by sufferers of arthritis, muscular dystrophy, diabetes, fibromyalgia, and many other painful conditions. This conclusion is reinforced by another study conducted by researchers in Spain. Wallace et al. (2007) injected volunteers with capsaicin (which is the active ingredient in Bengay cream). The result of doing this provides a particularly good model for pain. They then administered low, medium, and high doses of botanic cannabinoids and terpenes. They found that at low levels, there was essentially no pain relief; at moderate levels, there was excellent pain relief; and most significantly, at high levels, there was no pain relief or even increased pain.

The final study that needs to be addressed in relation to dosing and the endocannabinoid system comes from scientists in the United States (Hirvonen et al., 2012). The methodology and results of this study are fascinating because the scientists collected people that had been smoking 30 joints a day for years and so obviously had time to devote to an in-patient study. These are also people whose endocannabinoid system was well downregulated (out of balance) because they were ingesting so much THC. To comply with U.S. policy related to research on non-synthetic cannabinoids, the researchers were attempting to prove that botanic cannabinoid use causes brain damage because the only way to get permission to study biologically produced cannabinoids in America is by attempting to prove that they cause harm. This study essentially entailed locking the test subjects up for 28 days and not allowing them access to botanic cannabinoids. The result of the study was that after the 28 days were up, their endocannabinoid system returned to normal baseline levels in all the participants, as documented by brain scans. This study demonstrates that you can smoke 30 joints a day for years, stop for 28 days, and your endocannabinoid system essentially resets itself.

Chapter 7

**Using Cannabis Strain Fingerprints to
Target Specific Medical Conditions**

Albert Einstein is credited with the notion that if you can't explain your concept simply, you simply don't understand your concept. This chapter provides, in simple terms, a doctor of biomolecular psychology's understanding and first-hand experience with many of the chemovars of medicinal cannabis in a way that can be emulated by patients who have little experience and understanding of scientific methodology. My initial goal when I began medicating with cannabis was to control the neuropathic pain caused by my muscular dystrophy. Most of the strains I could obtain were successful at doing that, but I discovered that certain chemovars did other things, such as combatting depression, promoting focus, alleviating anxiety, and enhancing creativity. As a biomolecular psychologist, I was curious. How did these strain differences manifest themselves on a biomolecular level? I began to look at the research of the individual molecules and target these molecules to see how they reacted with my body chemistry.

When searching for a strain of cannabis to target a particular medical or psychological condition, it is essential that the patient knows what they're getting. This is why laboratory testing should never be overlooked. Testing facilities like Steep Hill Laboratories provide patients with a complete cannabinoid and terpene profile of their medication.

Despite the efforts of the United States as an institution to suppress studies of botanic cannabinoids by American scientists, research on the medicinal properties of these molecules has been ongoing in less scientifically repressed countries. In the 1970s, the United States had the power to suppress the pursuit of truth in other countries, but as its power dwindled, the research on the medicinal properties of *Cannabis sativa L* by scientists in these countries increased. This is the nature of scientific inquiry. As a survival mechanism, scientists have become conditioned to conform to the institutions that wield the power to determine which aspects of truth they are permitted to pursue. While researchers in the United States were prohibited from researching cannabinoid molecules unless they were produced synthetically, the U.S. NIH grants funded the research of scientists in less scientifically repressed countries to conduct the research that resulted in the discovery of the endocannabinoid system.

So, studies involving these nutraceutical molecules are available, and this chapter is designed to provide a useful way of using the laboratory analyses of cannabis chemovars to determine the appropriate strains for the medical results patients want to achieve. A laboratory report is referred to as a Certificate of Analysis (COA). In their roles as a functionary of the American Healthcare Delivery System, United States dispensaries sometimes resist providing these laboratory reports. Still, with enough patient persistence, these COAs are often attainable. Dispensaries are institutions, and institutions wield power. Some dispensaries push this power so far that they will not provide a certificate of analysis until after the product is purchased.

A Certificate of Analysis is an accredited laboratory document confirming that a regulated product meets certain specifications. A COA commonly contains the testing results performed as part of the quality control process. Most states with cannabis programs require laboratory testing on all products. The test results should be available on the company's website. If they are not, contact the company and request it. If they are unresponsive or report that the information isn't attainable, avoid purchasing from that dispensary.

COAs provide the molecular profiles of individual cannabis strains. This involves the botanic cannabinoid and terpene concentrations. These entail the major botanic cannabinoids of THC and CBD, as well as the minor botanic cannabinoids such as CBG, CBC, CBN, CBC, THCV, and CBL. Every analytical laboratory presents its COA somewhat differently, and some are easier to read than others. In the middle of the last decade, Steep Hills Laboratories placed into the public domain one of the most well-presented laboratory analyses of traditional and well-known strains of cannabis, and these provide illustrations of differences in chemovars of cannabis that are intuitively understandable. Steep Hills referred to their laboratory reports as Strain Fingerprints and provided the average concentrations of twelve of the most relevant cannabinoids and terpenoids found in cannabis strains. Below is the strain fingerprint of one of the most popular strains of medical cannabis in the nation. It is called Blue Dream and is widely referred to as the Bayer Aspirin of cannabis.

Be aware that there are thousands of strains of cannabis. Some of the strain names are strange and made up. Also, be aware that the name Google was made up, so don't be put off by the idiocy of some of these names. The key is the medicinal molecules the strain contains. The three main categories of compounds found in cannabis are cannabinoids, terpenes, and flavonoids.

Cannabinoids are compounds produced by the cannabis plants that, when ingested, interact with the body's endocannabinoid system, the system of receptors that control or influence all other bodily systems and play a vital role in the regulation of processes and functions, including sleep, pain, memory, and mood. There are more than 100 cannabinoids that can be found in cannabis, and most certified laboratories test for at least seven.

The cannabinoids mainly relieve the patient's physical symptoms but also have some mood-altering and enhancing properties The terpenes also have medicinal value but mainly seem to provide that sense of well-being for which cannabis is so well known. All plants produce terpenes. Most only produce one. Cannabis has been demonstrated to produce over 500 different terpenes. Flavonoids are not typically tested for but are highly diversified plant pigments that are present in strains of cannabis. Specific flavonoids have been identified to have a wide range of biological properties that can contribute to beneficial effects on human health. They are regularly consumed in the human diet and exhibit various biological activities, including anti-inflammatory, anti-cancer, and antiviral properties. Flavonoids also function as antioxidants, enzyme inhibitors, hormones, or immune system modulators. Flavonoids may be one of the safest non-immunogenic compounds because they are small organic molecules that stay active in the human body for a long time (Lee et al., 2007).

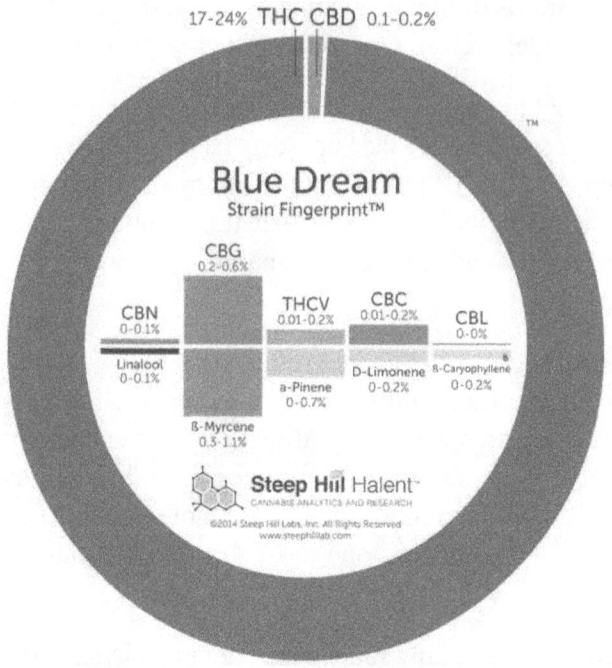

When I evaluate a strain to treat the neuropathic pain of muscular dystrophy, the first thing I consider is the ratio of the two major cannabinoids, THC and CBD. These ratios are located on the outer part of the ring. Generally, the first molecule I look for as a doctor of biomolecular psychology is not THC but rather CBD because the CBD molecule mitigates the intoxicating effects of THC. For muscular dystrophy pain, a strain with a high percentage of THC is recommended because it is the major neuropathic pain-killing molecule in cannabis. Patients experiencing neuropathic pain need the pain-relieving aspects of that molecule due to their medical condition. The intoxicative effect of THC is mitigated by choosing strains that have a high ratio of CBD. As has been discussed, moderate amounts are necessary. Essentially, the ratio of THC to CBD in the Blue Dream strain is appropriate enough to avoid the "high" but manage whatever pain issues individuals afflicted with muscular dystrophy might have. Still, this strain might produce a high because, like any other medication, it depends on the patient's body chemistry.

Although the medicinal molecules produced by plants have been used therapeutically since the Neolithic Era, these bullets were obliterated from the arsenal of Western medicine in 1906 because they cannot be patented. *Cannabis sativa L* has more medicinal molecules than any other plant, and these natural compounds have proven more efficacious than synthetically produced molecules because the latter may cause unwanted side effects due to the body's inability to degrade them. The side effects of cannabis intromission are typically positive. For example, a patient may be using botanic cannabinoids and terpenes to treat pain but also benefit from the side effects of the antibiotic, antiviral, antifungal, antimalarial, and antipathogenic properties of these nutraceutical molecules.

Major Cannabinoids

So, let's examine the medical benefits of THC and CBD, which are generally referred to as the major cannabinoids.

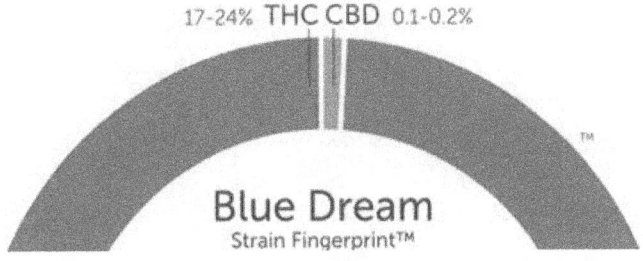

Cannabidiol (CBD)

Cannabidiol (CBD) is often the second most abundant cannabinoid in cannabis. It has serious implications in the field of medicine and is usually the most sought-after compound by medical users. It is a non-psychoactive component that mitigates the effect of THC. This means that strains high in both THC and CBD will induce much clearer head highs than hazier heady strains that contain low amounts of CBD and huge amounts of THC. CBD has a long list of medicinal properties, the most significant of which is that it relieves such things as chronic pain, inflammation, migraines, arthritis spasms, Crohn's disease, epilepsy, anxiety, and schizophrenia. CBD has also been shown to have antibacterial, antiviral, and anti-cancer properties, and new medicinal uses are being found all the time as more research is conducted on the medicinal properties of the CBD molecule (Peng et al., 2022).

Δ9-Tetrahydrocannabinol (THC)

Delta-9- tetrahydrocannabinol (THC) is the most abundant cannabinoid in most strains of cannabis and is claimed to be responsible for the main bliss-inducing effect experienced upon cannabis consumption. This is because it activates the CB1 receptor, the receptor responsible for dopamine release, creating a sense of euphoria and feelings of well-being. THC also has analgesic effects, relieving the symptoms of neuropathic pain and inflammation. Combined with other botanic cannabinoids, this causes a great feeling of relaxation. Additionally, THC has antibacterial properties. It shares this characteristic with the other botanic cannabinoids described in this chapter (Appendino et al., 2008).

Minor Cannabinoids

After the patient determines the best ratio of THC to CBD for their body chemistry, it is time to start to look at what are referred to as minor cannabinoids.

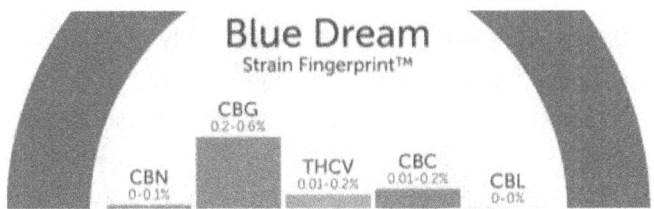

Minor cannabinoids will become more and more major as the research progresses. A typical strain fingerprint contains six cannabinoids (CBN, CBG, THCV, CBC, and CBL). We'll examine them in the order they are presented on the strain fingerprint.

Cannabinol (CBN)

Cannabinol (CBN) is purported to be the main cannabinoid in cannabis that makes you sleepy. Many medical cannabis patients struggle with insomnia and so tend to look for a strain that is high in ratio of CBN. These are difficult to find because strains that are high in CBN are rare. CBN is produced when THC oxidizes (breaks down) the result of being stored for long periods of time. Most medical cannabis patients don't want to spend the time or be troubled with oxidizing their medicine but still have a hard time finding strains with ratios high in CBN at dispensaries.

As it happens, cannabis that has been seized by law enforcement and sits in evidence lockers for months on end turns out to have high levels of CBN when laboratory tested, but for obvious reasons, most medical cannabis patients hesitate to go to the police and ask for a sample of their stash. As already mentioned, cannabis is widely used as a sleep aid for those who suffer from insomnia, and CBN is purported to be one of the reasons why. Even so, there is insufficient published evidence to support sleep-related claims. Randomized controlled trials are still needed to substantiate claims made by manufacturers about cannabis products containing CBN. These studies should specifically evaluate its effects on sleep through polysomnography, or at minimum, through validated sleep questionnaires, and use dosages significantly higher than those found in currently available cannabis products marketed for sleep. Individuals seeking cannabinol-derived sleep aids should be skeptical of manufacturers' claims of its sleep-promoting effects.

A substantiated use for cannabinol is as an antibacterial. According to an Italian study in 2008, cannabinol showed potent activity against MRSA when applied as a topical (Appendino et al., 2008). Topical uses have also shown promise in treating burns and psoriasis (Baswan et al.,2020). The research on CBN is preliminary, but studies suggest it could stimulate bone cell growth (Fellous et al., 2020; Scutt & Williamson, 2007). If that's the case, it would prove beneficial in the treatment of osteoporosis. It could also help those with broken bones recover more quickly.

Cannabigerol (CBG)

CBG, or cannabigerol, is one of the hundreds of minor cannabinoids found in cannabis and hemp plants. It is referred to as the "mother of all cannabinoids" because it is the precursor of other cannabinoids, such as THC, CBD, CBC, and CBN (Russo et al., 2022, p. 707). Cannabigerol does not activate dopamine, and so is a non-intoxicating botanic cannabinoid. Still, it exhibits multiple therapeutic properties, including antibacterial, antifungal, anti-inflammatory effects, and anti-cancer properties (Borrelli et al., 2013; Lah et al., 2021; Cabrera et al., 2021). It has also been demonstrated to stimulate brain and bone cell growth (Calapai et al., 2022; Fellous et al., 2020). CBG is also one of the cannabinoids in cannabis that treats inflammatory bowel disease (Kogan et al., 2021; Nudma et al., 2023). Similar to other botanic cannabinoids, cannabigerol's complex biological effects are believed to result from modifications of dependent redox and inflammatory processes, which modulate cellular metabolism.

A survey study examining the perceptions of patients who reported using CBG-predominant cannabis chemovars was conducted by Russo et al. (2022) and revealed that the most common conditions the patients reported using CBG to treat were anxiety (51.2%), chronic pain (40.9%), depression (33.1%), and insomnia or disturbed sleep (30.7%). Efficacy was highly rated, with the majority reporting that their conditions were "very much improved" or "much improved" by intromitting CBG. Nearly three-quarters (73.9%) of the participants asserted the superiority of CBG-predominant cannabis strains over conventional medicines for chronic pain, 80% for depression, 73% for insomnia, and 78.3% for anxiety.

While survey studies have scientific value, the gold standard for research on the mood-enhancing properties of this molecule would entail a double-blind, placebo-controlled cross-over trial to examine the acute effects of CBG on anxiety, stress, and mood. The results of that study were released as this book was going to the publisher. Cuttler et al. (2024) released the results of a double-blind, placebo-controlled cross-over field trial to examine the acute effects of CBG on anxiety, stress, and mood. The results of this double-blind, placebo-controlled, cross-over field trial indicated that 20 mg of hemp-derived CBG reduced patient anxiety and stress in healthy cannabis-using adults sans motor or cognitive impairment, intoxication, or other subjective side effects that are typical of pharmaceutical medicines designed to treat anxiety, stress, and mood disorders.

Cannabigerol (CBG) is interesting because it synergizes the medicinal properties of the other botanic cannabinoids. Another interesting aspect of cannabigerol is that it inhibits the uptake of GABA. GABA is a brain chemical that determines how much stimulation a neuron needs to cause a reaction. When GABA is inhibited, it decreases anxiety and muscle tension, similar to the effects of CBD (Russo et al., 2022). Along with CBD, it is one of the botanic cannabinoids that treat glaucoma because CBG increases fluid drainage from the eye and reduces the amount of pressure (Calapai et al., 2022; Kogan et al., 2021). CBG also has antidepressant properties and inhibits tumor growth (Bęben et al., 2024; Nachnani et al., 2021).

Tetrahydrocannabivarin (THCV)

Tetrahydrocannabivarin (THCV) is a compound in cannabis that offers a unique array of effects and medical benefits that sets it apart from other botanic cannabinoids. THCV is likely to become significant in the relatively near future because, unlike other cannabinoids, it dulls the appetite. This may be good for patients focused on weight loss but should be avoided by people battling anorexia or lack of appetite. Furthermore, this minor cannabinoid is going to become major soon because of its ability to regulate blood sugar levels and reduce insulin resistance. This means that it has real potential as a possible treatment for diabetes. THCV decreases appetite, increases satiety, and up-regulates energy metabolism, making it a clinically useful remedy for weight loss and management of obesity and type 2 diabetes. The effect of THCV on dyslipidemia and glycemic control in type 2 diabetics demonstrated reduced fasting plasma glucose concentration when compared to a placebo group. Furthermore, the uniquely diverse properties of THCV provide neuroprotection, appetite suppression, and glycemic control, with positive side effects, making it a potential priority candidate for the creation of clinically useful future therapies while providing a platform for the treatment of life-threatening diseases (Abioye et al., 2020).

Additionally, THCV reduces panic attacks and curbs anxiety attacks in PTSD patients without suppressing emotion (Ahmed et al., 2022). This minor botanic cannabinoid may be another of the many that help with Alzheimer's disease. Tremors and motor control issues associated with Alzheimer's appear to be improved by THCV, but research is ongoing (Stone et al., 2020). THCV is yet another cannabinoid that stimulates bone growth. Because it promotes the growth of new bone cells, THCV is being looked at as a treatment for osteoporosis and other bone-related conditions (Fellous et al., 2020).

Cannabichromene (CBC)

Cannabichromene, sometimes the second most abundant compound in cannabis, is non-intoxicating and has also been shown to relieve pain and inflammation. CBC activates CB2 receptors, but not CB1 receptors and therefore is a selective CB2 receptor agonist that can be utilized for CB2 receptor-related regulation of inflammation. Furthermore, TRPV receptors and PPAR receptors represent prominent molecular targets of CBC. There is also the possibility that, like other botanic cannabinoids, CBC may exhibit an affinity for additional receptors. The diverse range of receptor affinities exhibited by CBC underlies its multifaceted pharmacological attributes, suggesting the potential usefulness of this botanic cannabinoid as a viable compound capable of eliciting not only anti-inflammatory effects but concurrently a spectrum of other effects, including the treatment of migraines and the stimulation of bone cell growth (Apostu et al., 2019; Baron, 2018; Hong et al., 2023).

A 2013 study by Shinjyo and Di Marzo suggested that CBC increases the viability of developing brain cells (neurogenesis). This was substantiated in 2023 by a study conducted by Valeri et al. (2023). Contrary to popular belief, neurogenesis doesn't stop once you reach a certain age. However, it only occurs in a specific part of the brain known as the hippocampus. The hippocampus is important for memory and learning, and the lack of growth in this area is believed to contribute to several disorders, including depression and Alzheimer's. While cannabicromene's ability to promote neurogenesis is a very recent finding, previous studies suggest THC and CBD do the same (Jiang et al., 2005; Prenderville et al., 2015; Wolf et al., 2010).

One of the first scientists to uncover this remarkable aspect of cannabis – Dr. Xia Jiang explained why in an interview with Science Daily: *"Most 'drugs of abuse' suppress neurogenesis. Opiates, alcohol, nicotine, and cocaine are all known to inhibit brain growth. Thankfully, CBC and other chemicals in marijuana seem to have the opposite effect."*

A study conducted by El-Alfy et al. (2010) identified a significant antidepressant effect of cannabichromene in rat models. The study revealed that CBC and some other cannabinoids may contribute to the overall mood-elevating properties of cannabis. Scientists are still trying to determine how CBC does this since it doesn't seem to activate the same neural pathways as THC.

Cannabicyclol (CBL)

Cannabicyclol

Studies of Cannabicyclol (CBL) are in their early stages. A few in vivo studies occurred in the 70s, and there are a few newer bench studies, but no human trials of its effects have been done. CBL has the right molecular structure to be an anti-inflammatory, but the concentration of CBL needed to stop inflammatory prostaglandins is reportedly less than other botanic cannabinoids tested in the same experiment (Burstein et al., 1973). An in vitro study suggests that CBL has anti-proliferative activity on human colon cancer cells, indicating that this botanic cannabinoid could be a promising agent for the prevention of human colorectal cancer (Lee et al., 2022). While general antibiotic potential is indicated, CBL demonstrated almost no inhibition of MRSA (Moloney, 2016). Comparable botanic cannabinoids are known for exhibiting anticonvulsant, vasorelaxant, pain-relieving, antianxiety, antiemetic, and neuroprotective properties (Klahn, 2020), but the degree to which CBL shares these benefits is yet to be determined.

There are a couple of other botanic cannabinoids that require mentioning at this point, and it's unfortunate that they aren't included in a strain fingerprint, but there are over 100 other cannabinoids, and they had to stop somewhere. So, let's examine THCA and CBDV as well. Many COAs include these botanic cannabinoids, and they do so for a reason.

Tetrahydrocannabinolic Acid (THCA)

Δ^9 -Tetrahydrocannabinolic Acid

(Δ^9 THCA)

Tetrahydrocannabinolic acid (THCA) is the acidic form of THC. As a cannabis plant matures and its buds grow, its cannabinoid and terpene profile begin to develop. The first cannabinoid the plant will create is CBGA, which will ultimately break down and produce primary cannabinoids like THCA and CBDA.

THCA is not intoxicating if ingested. Structurally, it has an additional molecular carboxyl ring, which prevents it from binding to receptors in the brain responsible for the "high feeling. When THCA is exposed to heat, such as when smoking, vaping, dabbing, or cooking, it will convert into the intoxicating botanic cannabinoid THC. This is referred to as decarboxylation, and this conversion alters the molecular structure of THCA by removing a carboxyl ring. The removal of this ring helps THC bind to the CB1 receptors in our bodies. Decarboxylating accentuates the release of dopamine, which results in the pleasurable sensation that is referred to as a "high."

Raw cannabis does not produce a high if consumed without decarboxylating first because the molecular structure of THCA prevents it from binding to the CB1 receptor (Lewis-Baker et al., 2019). Still, the tetrahydrocannabinol acid (THCA) compound has several therapeutic uses. This botanic cannabinoid works to relieve inflammatory pain (Henshaw et al., 2021) and is an ideal cannabinoid for treating symptoms of such conditions as arthritis and seizures (Benson et al., 2022). THCA is also an effective neuroprotectant (Nadal et al., 2017; Stone et al., 2020), so it is beneficial in the treatment of such conditions as multiple sclerosis, Alzheimer's, and Parkinson's disease (Kim et al., 2023). It can also help to stimulate the appetite in patients suffering from cachexia and anorexia nervosa (Lowe et al., 2021). Most impressively, research shows that THCA helps to slow the proliferation of cancerous cells (De Petrocellis et al., 2013). Furthermore, THCA is an antiemetic (Rock et al., 2021), suppressing nausea and vomiting. This botanic cannabinoid has also been demonstrated to aid with sleep (Russo, 2018), and THCA has been demonstrated to both improve and suppress immune functions (Verhoeckx et al., 2006). Suppression of the immune system is important at times because if the immune system gets out of control, it can mistake healthy cells for invaders and repeatedly attack them. These are generally referred to as auto-immune disorders (again, balance is tthe key). This is what happens with any inflammatory condition.

Cannabidivarin (CBDV)

CANNABIDIVARIN (CBDV)

It has become general knowledge that CBD demonstrates efficacy in treating certain forms of epilepsy. This has been demonstrated particularly effectively in some cases of epilepsy in children. There is a small percentage of these patients that do not respond to CBD as treatment. It is likely that many of these can be treated effectively with the cannabidivarin (CBDV) molecule. The problem is that the CBDV molecule is rare, and smart cultivators are beginning to target this particular botanic cannabinoid. The way it works is based on the fact that cannabis plants are heterozygous. That simply means that if you throw down a few seeds, you get a few types of medicinal molecules in your plants, but if you throw down a lot of seeds, you get a variety of medicinal molecules, and you can breed for the molecules you want. In the relatively near future, cultivators are going to be producing strains that are high in CBDV to modulate epileptiform activity in patients who do not respond to CBD as a treatment. In a similar fashion to cultivators breeding cannabis chemovars towards the molecules that they want to produce, medical cannabis patients can target strains for the medicinal molecules to treat their individual medical conditions. This benefits the patient by allowing them autonomy and provides an avenue for them to take control of their own healthcare.

Terpenes

So, those are the botanic cannabinoids, and as you can see, they have medicinal benefits a-plenty. However, as most medical cannabis patients know, most of the strain differences come from another group of medicinal molecules known as the terpenes. Virtually everyone is familiar with terpenes because every person has at one time or another, walked through a pine grove or smelled a rose. All plants have terpenes, and terpenes are the molecules in plants that give them their scent. Most plants have just a single terpene, but cannabis has hundreds, and many medical cannabis patients choose strains based on their terpene content. This is because it is the terpenes in cannabis that improve a person's perception of their own well-being. Terpenes are generally considered by cannabis patients to be a state-of-mind medication.

While it is true that it is likely the combined effect of the terpenes that provide the patient the sense of well-being that cannabis is so well known for, terpenes have some medicinal value. If you know anything about holistic medicine, you are probably familiar with some of the terpenes. Sometimes, they are referred to as essential oils, but they all come from plants. Most plants have a single terpene, cannabis has hundreds, and many seem to have a broad range of biological properties, including analgesics, cancer chemo-preventive effects, antimicrobial, antifungal, antiviral, antipathogenic, anti-hyperglycemic, antimutagenic, anti-inflammatory, anti-parasitic activities, and memory enhancers. (Cox-Georgian et al., 2019; Dash et al., 2022).

These medicinal benefits are virtually ignored by most medical cannabis patients because many probably feel they get all that they need from the cannabinoids. Still, the entourage effect suggests that botanic cannabinoids and terpenes work together to enhance a person's well-being. This will become clear as we examine the terpenes in a strain fingerprint more closely. This is Strain Fingerprint of another very popular strain of cannabis that is well known for its terpene content:

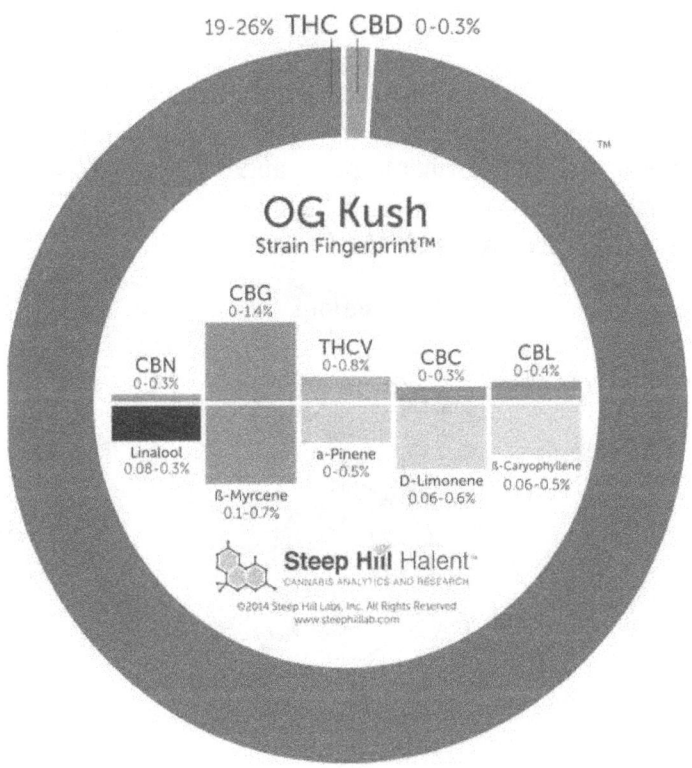

Myrcene

Myrcene is also found in mangos and is good for insomnia. It is probably the most interesting of the terpenes because it synergizes the antibiotic properties of the other terpenes and it changes the permeability of the cell membranes to allow for better absorption of the cannabinoids into the brain (Sieniawska et al., 2017).

Linalool

Linalool is found in the lavender plant and exhibits multiple bioactive properties, including anticancer, antimicrobial, neuroprotective, anxiolytic, antidepressant, anti-stress, hepatoprotective, renal protective, and lung protective activity. Besides this, linalool can induce apoptosis of cancer cells *via* oxidative stress while concurrently protecting normal cells. Linalool exerts antimicrobial effects through the disruption of cell membranes. The protective effects of linalool to the liver, kidney and lung are owing to its anti-inflammatory activity. Because of its protective properties and low toxicity, linalool can be used as an adjunct to anticancer drugs or antibiotics. These properties suggest that linalool has great potential to be a natural and safe alternative therapeutic (An et al., 2021).

Limonene

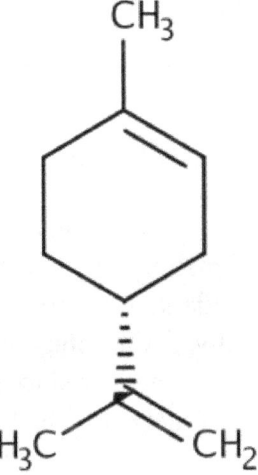

Limonene is found in lemons and other acidic fruits. The therapeutic effects of this terpene have been extensively researched, and studies revealed anti-inflammatory, antioxidant, antinociceptive, anticancer, antidiabetic, antihyperalgesic, antiviral, and gastroprotective effects, among other beneficial effects in the maintenance of health (Vieira et al., 2018). In a recent and significant study conducted in cooperation with Johns Hopkins University School of Medicine (Spindle et al., 2024), Δ-Limonene selectively attenuated the anxiety-inducing effects of THC.

This serves as confirmation of the entourage effect and indicates that this terpenoid may increase the therapeutic index of THC.

β-Caryophyllene

β-Caryophyllene is found in many plant foods, including oregano, black pepper, and clove. This terpene possesses diverse therapeutic properties, including antioxidant, anti-inflammatory, antidiabetic, and analgesic effects. Additionally, β-Caryophyllene exerts antioxidant and neuroprotective effects that can be utilized to treat neurological diseases and mood disorders such as anxiety (Machado et al., 2020).

Pinene

α- and β-pinene are mainly produced by pine trees and other conifers, in addition to a variety of herbs such as rosemary, parsley, basil, and even the peels of oranges (Erman & Kane, 2008; Vespermann et al., 2017). Pinene has been demonstrated to boost focus (Weston-Green et al., 2021), and studies reveal that pinene may have anti-inflammatory properties, making it a potentially valuable compound for managing inflammatory conditions (Kim et al., 2015). Additionally, pinene has been demonstrated to be an anti-cancer agent, inhibiting the growth of cancer cells and inducing programmed cell death (Zhang et al., 2015). Psychologically, inhaling pinene-rich essential oils enhances mood, promotes calmness, and reduces stress and anxiety (Vora et al., 2024).

The most effective method of determining the medicinal molecules you need is to have a doctor diagnose your medical condition. Once your medical condition is identified, you can use the following wheel to determine the botanic cannabinoids to target that ailment. You can then use the strain fingerprints to find strains of cannabis that are heavy in those molecules. Notice that it is quite rare to find a single molecule alone.

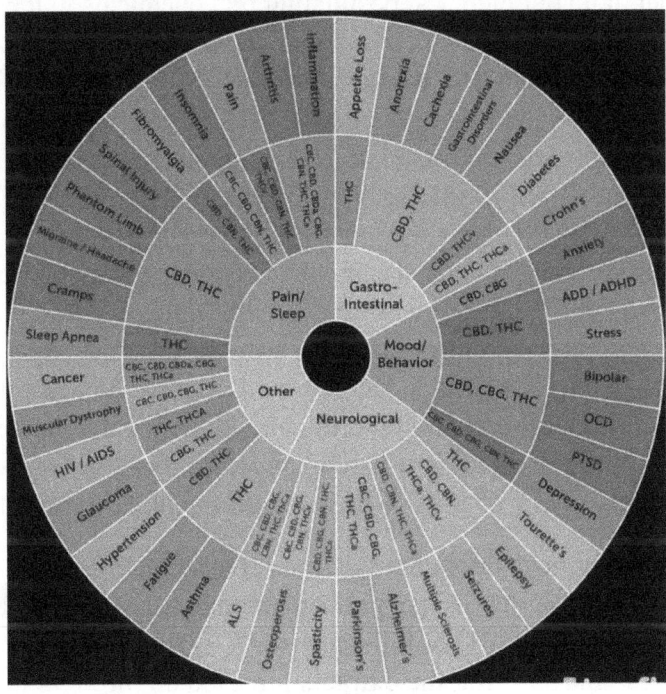

The Entourage Effect

Until recently, perhaps the most controversial construct involving cannabinoid-based therapies has been the entourage effect, the idea that constituents in cannabis chemovars act synergically to magnify or mitigate the supplement's effects. Dr. Raphael Mechoulam, a biochemist that NIH funded at the Hebrew University of Jerusalem, was the first researcher to propose the entourage effect theory when he discovered that 2-linoleoylglycerol, 2-oleoylglycerol, and 2-palmitoylglycerol do not bind to cannabinoid receptors but enhances the binding potentiation of 2-arachidonoylglycerol (Ben-Shabat et al., 1998). This observation inspired Mechoulam to propose that the congeners synergistically enhance the endocannabinoid's binding potential to the primary cannabinoid receptors, theoretically through inhibition of 2-arachidonoylglycerol (Murataeva et al., 2016).

Ethan Russo, a neurologist and former director of research and development at the International Cannabis and Cannabinoids Institute in Prague, extended the entourage effect theory to the theory of endocannabinoid deficiencies, postulating that chemicals in the cannabis plant could enhance, heighten, or mitigate the therapeutic effects of THC. Russo contended that CBD works to enhance THC's therapeutic effects (Russo, 2011). As evidence, he cited a 2010 clinical trial of Sativex, a botanical compound comprised of THC and CBD, to treat neuropathic pain in people with multiple sclerosis (Serpell et al., 2014). The study consisted of 177 participants and had three arms. The first group received a placebo, the second was given a compound containing high concentrations of THC, and the final group was treated with Sativex. Participants were asked to score their pain throughout the two-week clinical trial and state at the end how much their pain had lessened, if at all. A reduction in pain of 30% or more was considered clinically significant. Approximately 40% of the people treated with Sativex reported this level of pain relief, almost twice as many as those who received the placebo or THC alone.

Another study supporting the existence of an entourage effect is a 2018 meta-analysis involving 670 people with treatment-resistant epilepsy and given either purified CBD or full-spectrum CBD-rich cannabis extracts. 71% of those intromitting the extracts reported an improvement in the frequency of seizures, compared with 46% of people given purified CBD (Pamplona et al., 2018).

The entourage effect theory has been seized upon by cannabis suppliers and is promoted relentlessly as a marketing tool for products containing a full spectrum of botanic cannabinoids and terpenes. As with many scientific theories, the entourage effect was not universally accepted within the scientific community, and a researcher's acceptance of the construct appears to be correlated to whether they have been propagandized by the nutraceutical or pharmaceutical paradigms (Hesselink, 2013). The pharmaceutical industry is mired in a one-molecule approach to medicine and has funded studies that assert the synergizing components are not inherently pharmacologically active, suggesting the construct is merely a marketing ploy by marijuana companies peddling illicit drugs with no federally accepted medicinal components. Scientists adhering to both paradigms exhibit some bias in their publications, and objective researchers must constantly scrutinize the motivations behind how and why scientific knowledge is constructed.

Recently, a group of researchers replicated the experiment in which Mechoulam and his colleagues studied the endocannabinoid 2-arachidonoylglycerol (2-AG), which binds to primary cannabinoid receptors (Ferber et al., 2020). As previously discussed, the Mechoulam group discovered that in mice's brains, spleens, and guts, 2-AG is characteristically found together with two other compounds: 2-linoleoylglycerol and 2-palmitoylglycerol. Unable to activate the primary CB1 and CB2 receptors themselves, these two molecules facilitate 2-AG's potential to bind to the receptors and increase effects such as analgesia in the animals.

The study, conducted in 2016, replicated the Mechoulam experiment, examining the unknown but closely related lipid species with fatty acids of different lengths and saturation: 2-oleoylglycerol, 2-linoleoylglycerol, and 2-palmitoylglycerol. This replication utilized cutting-edge instrumentation to examine whether these lipid progenitors are degraded by the same enzymes as 2AG, thereby competing with 2-AG for breakdown. If this competition exists, the result would be an enhancement of 2-AG concentrations and a prolongation of 2-AG signaling, sans an entourage effect (Murataeva et al., 2016). To test this proposition, the researchers replicated the experiment conducted by Mechoulam using instrumentation to which Mechoulam had no access, attempting to dispute whether the progenitors he and his group didn't know existed act in a manner inconsistent with the role of entourage compounds in these diverse models of 2-AG signaling. They had access to better cell lines than the immortalized cell lines the Mechoulam group used and did not fully recapitulate endogenous cannabinoid signaling. Instead, they utilized autaptic hippocampal neurons, a model system that possesses the necessary mechanism to suppress activation.

Using a superior experimental design, advanced technology, and better cell lines, the results of the replication of the Mechoulam experiment showed that 2-oleoylglycerol, 2-linoleoylglycerol, and 2-palmitoylglycerol do not behave in a manner consistent with entourage compounds. Instead, all compounds, but most notably 2-oleoylglycerol, acted as antagonists. 2-palmitoylglycerol very slightly antagonized the CB2 receptors, while 2-oleoylglycerol demonstrated the most significant antagonistic effect in neurons and weakly antagonized the CB1 receptors. While this replication study was well designed and analyzed fatty acids of which the Mechoulam group was unaware and incorporated instrumentation and cell lines to which they had no access, the results as reported fail to negate the fact that endocannabinoid system compounds act synergically to modulate the effects of the endocannabinoids on the primary and secondary cannabinoid receptors. The fact that 2-oleoylglycerol and palmitoylglycerol act as antagonists indicates they mitigate the effects of 2AG and is indicative of an entourage effect. Additionally, the researchers point out that 2-oleoylglycerol, 2-linoleoylglycerol, and 2-palmitoylglycerol may independently activate the GP55 receptors, with no entourage effect involved. There are multiple claims in the literature that the GP55 receptors are cannabinoid receptors because most biologic and synthetic cannabinoid molecules activate them (Dalton et al., 2009; Nevalainen & Irving, 2010; Shahbazi et al., 2020; Yang et al., 2016). Respect for this research wanes because few molecules seemingly do not turn the GP55 receptors on (Rahimi et al., 2015). Furthermore, even if no entourage effect existed in this instance, it would not eliminate its existence in other instances.

Recently, a study was conducted that indisputably demonstrated the existence of the entourage effect of botanic cannabinoids and terpenes

(Spindle et al., 2024). The research was conducted in association with The Behavioral Pharmacology Research Unit of the Department of Psychiatry & Behavioral Sciences at Johns Hopkins University School of Medicine. Ironically, it was funded by a National Institute on Drug Abuse (NIDA) research grant. It utilized a double-blind, within-subjects crossover design, which is considered the gold standard for medical research. The study methodology entailed twenty healthy adults completing nine double-blind outpatient sessions in which they inhaled vaporized THC alone (15 mg or 30 mg), Δ-Limonene alone (1 mg or 5 mg), the same doses of THC and Δ-Limonene together, or a placebo. Twelve participants completed a tenth session in which 30 mg THC+15 mg Δ-Limonene was administered. Pharmacodynamic outcomes included subjective drug effects, cognitive/psychomotor performance, vital signs, and plasma THC and Δ-Limonene concentrations.

When Δ-Limonene was administered alone, pharmacodynamic outcomes did not differ from placebo. Administration of 15 mg and 30 mg THC alone produced subjective, cognitive, and physiological effects typical of acute cannabis exposure. Ratings of anxiety-like subjective effects qualitatively decreased as Δ-Limonene dose increased, and concurrent administration of 30 mg THC+15 mg Δ-Limonene significantly reduced ratings of "anxious/nervous" and "paranoid" compared with 30 mg THC alone. The results of this study indisputably demonstrate an entourage effect and that the money the pharmaceutical industry pays scientists to publish studies disputing its existence does nothing to change the fact that it is a natural phenomenon.

The pharmaceutical industry has introduced an array of synthetic drugs aimed at manipulating the immune system and treating the effects of acute inflammation. The most prescribed medications by the American Healthcare Delivery System are corticosteroids and Non-steroidal Anti-inflammatory Drugs (NSAIDs). Despite being the universal first choice in treating inflammatory diseases, these drug classes have become known for their adverse side effects, limiting their dosage and prolonged treatment regimens (Heffler et al., 2018; Kim & Solomon, 2011; Sarnes et al., 2011; Wehling, 2014).

Studies reveal that, unlike synthetic drugs, the therapeutic advantage of medicinal cannabis is attributed to the combined mechanisms of blended compounds as the result of synergisms or antagonisms (Mahmud et al., 2021). Synergy may occur among cannabinoids (intra-entourage) or between cannabinoids and terpenes (inter-entourage) (Blasco-Benito et al., 2018; Vermeij et al., 2014). Terpenes and flavonoids play essential roles in modulating cannabinoid functional ability by altering pharmacokinetics and permeability. They can either increase therapeutic activity or decrease toxicity by interacting with various cellular and physiological systems (Nahtigal et al., 2016; Stasiłowicz et al., 2021).

Pharmaceutical companies eager to create synthetic cannabinoid medicines are always looking to synthesize one or two patentable molecules. Marinol, sold in the U.S. by AbVie, is a pill containing synthetic THC that's used to stimulate appetite in cancer and HIV patients. GW Pharmaceuticals has patented Sativex spray, a cannabinoid product made from a mixture of synthetic THC (Nabilone) and cannabidiol, to treat pain in multiple sclerosis patients (MacDonald & Adams, 2019). But the problem with these synthetic cannabinoids is that they don't work as well as whole plant therapy, and a small bottle of Sativex spray costs about sixteen thousand dollars. The pharmaceutical company Jazz owns the patent on Epidiolex, a concoction of sucralose and cannabidiol designed to alleviate seizures associated with Lennox-Gastaut syndrome. Sucralose functions as a strawberry-flavored artificial sweetener that makes the chemical palatable for children. More importantly, this synthetic component makes the drug patentable and able to achieve FDA approval. The side effects of Epidiolex listed in the results of the randomized, double-blind, placebo-controlled trials that led to its approval by the FDA were delineated by Thiele et al. (2018) as diarrhea, vomiting, acute hepatic failure, hepatic failure, viral infection, increased concentration of another antiepileptic drug in the blood, convulsion, lethargy, restlessness, acute respiratory distress syndrome, acute respiratory failure, hypercapnia, hypoxia, pneumonia aspiration, and rash.

As already mentioned, Marinol is a prescription drug that contains synthetic THC. Marinol was approved by the FDA in 1985 for treating the side effects of chemotherapy. However, many doctors and patients soon discovered that Marinol is a poor substitute for whole-plant therapy.

The most likely reason is directly related to the entourage effect, as Dr. Sanjay Gupta explains:

> When the drug became available in the mid-1980s, scientists thought it would have the same effect as the whole cannabis plant. However, it soon became clear that most patients preferred the whole plant to ingesting Marinol. Researchers began to realize that other components, such as CBD, might have a larger role than previously realized.

First described in 1998 by Israeli scientists Shimon Ben-Shabat, Raphael Mechoulam, and others, the basic idea of the entourage effect is that the medicinal cannabinoids work together, or produce synergy, and affect the body in a mechanism similar to the body's endocannabinoid system (Ben-Shabat et al., 1998).

This theory is the foundation for the idea that, in certain cases, whole plant extractions are better therapeutic agents than individual cannabinoid extractions. The entourage effect theory has been expanded in recent times by Wagner and Ulrich-Merzenich (2009) to include the four basic mechanisms of whole plant extract synergy as follows:

1. Ability to affect multiple targets within the body.

Many studies have demonstrated the effectiveness of cannabis as a therapeutic agent for inflammation throughout the body. Researchers attribute this to the presence of cannabinoids which help facilitate the activity of the body's endocannabinoid system.

2. Ability to improve absorption of active ingredients.

The entourage effect can also work to improve the absorption of cannabinoids. Cannabinoids are chemically polar compounds, which make them, at times, difficult for the body to absorb in isolation. Absorption of topicals provides a prototypical example of this problem. The skin is made up of two layers, also known as a bilayer, which makes it difficult for very polar molecules like water and cannabinoids to pass through. However, in addition to their own anti-inflammatory properties, terpenoids, like beta-caryophyllene, assist in the absorption of cannabinoids, increasing the therapeutic benefits of the botanic cannabinoids (Blake, 2021; Ricardi et al., 2024).

3. Ability to overcome bacterial defense mechanisms.

The entourage effect also accounts for cannabis being effective in treating various bacterial infections. There are various studies that prove the antibacterial properties of botanic cannabinoids (Gildea et al., 2022; Klahn, 2020; Nidir et al., 2020), and terpenoid constituents of cannabis also have potent antibacterial properties as well with α-pinene, β-pinene, β-myrcene, and limonene being the most effective (Iseppi et al., 2019; Jokić et al., 2022).

4. Ability to minimize adverse side effects.

Finally, the entourage effect allows certain cannabinoids to modulate the negative side effects of others. The most fitting example of this is CBD's ability to modulate the perceived negative effects of THC (Szkudlarek et al., 2021). Many patients have heard about (or experienced) the increased anxiety and paranoia sometimes associated with cannabis overconsumption. Thanks to the entourage effect, research has shown that CBD can be effective in minimizing the anxiety associated with THC, lowering the user's feelings of paranoia. As research progresses, it becomes evident that THC, CBD, and the remaining cannabinoids do not compete with each other but act in tandem alongside other cannabis components to provide therapeutic relief for a wide variety of ailments (Spindle et al., 2024).

The Entourage Effect and Its Relationship to Creativity

There is little doubt that in people who inherently have creative talent, cannabis tends to increase that ability (Kowal et al., 2015). Creativity is a concept that is difficult to explain. It's kind of like art; you know it when you see it. The reason cannabis enhances creativity must be directly related to the entourage effect. It's important to understand that there is no possible way that creativity can come from just one area of the brain. Creativity, by definition, must come from all parts of the brain, and all the different areas of the brain must work together to enhance imagination. Creativity and imagination would have to have such a vast array of chemical components that we will never be able to analyze them all on a biomolecular level. However, we understand certain components of creativity and can look at those components from a molecular perspective. For example, one of the most relevant and easiest to study components of creativity is focus; the ability to stay on task and not be distracted by outside events. We can see what region of the brain is stimulated when someone is focused on a task, and pinene stimulates that portion of the brain (Hashemi & Ahmadi, 2023).

Creativity can't be looked at in the same way. You can't look at creativity the same way you look at focus. Creativity is too complex. The other components of creativity are not that easy to identify, but there are certainly other aspects of imagination that tend to be activated on a biomolecular level. There are currently over 500 identified terpenes in cannabis, and in the same way the botanic cannabinoids all work together to make a person feel better, the terpenes all work together to enhance a person's creative abilities. This entourage of the terpenes working together to enhance creativity is so similar to the entourage of the cannabinoids working together to relieve physical symptoms it's eerie.

The Medicines in Cannabis

So, the question becomes, where do you find the medicine in the cannabis plant? For the most part, you only really find medicine when the female plant matures and produces buds. This is a bud on a mature OG Kush plant.

This is what it looks like dried.

Notice it looks kind of hairy. These hairs are called trichomes. This is what trichomes look like microscopically.

The little round balls on the top are where the medicine is located, and if you collect those balls, you get what is called kief, which is useful in the making of edibles.

Kief

If you press those balls together, you get what is known as a hash, which, as Jeb Bush discovered in prep school, is useful for other things (Jeb Bush was a hash dealer in prep school). It's not slander if it's true (I checked).

Hash

Summary

The point is that many of the medicinal molecules produced by the cannabis plant and their therapeutic properties have been individually explained, and this information can be used by clinicians and patients to target specific mood, autoimmune, and endocannabinoid deficiency disorders. The following list is intended to provide a simple method to select molecules to target specific medical goals:

- Analgesic (Relieves pain)-CBC, CBD, THC, THCA THCV, CBL, CBN, Limonene, Myrcene

- Antidiabetic (Reduces blood sugar levels) CBD, THCV, CBG.

- Antibacterial (Inhibits bacterial growth) CBD, CBN, CBG, CBC, CBC, Limonene, Myrcene, Pinene, Linalool, Beta-Caryophyllene.

- Antidepressant (Relieves symptoms of depression) THC, CBC, CBG, CBC, CBN, Limonene, Myrcene.

- Anti-emetic (Reduces nausea and vomiting) THC, CBD.

- Anti-epileptic (Reduces seizures and convulsions) THCA, THCV, CBN, CBD, CBDV, Linalool.

- Anti-fungal (Treats fungal infection) CBC, CBD, limonene, Beta-Caryophyllene, Eucalyptol

- Anti-inflammatory (Reduces inflammation) CBDA, CBD, CBC, CBCA, CBG, CBN, THC, THCA, Myrcene, Beta-Caryophyllene.

- Anti-insomnia (Aids sleep) CBN, THC, THCA, CBG Myrcene, Linalool

- Anti-ischemic (Reduces risk of artery blockage) CBD.

- Anti-proliferative (Inhibits cancer cell growth) THCA, CBC, CBCA, CBD, CBG, CBDA, THC, THCA, Limonene.

- Anti-psoriatic (Treats psoriasis) CBD, CBC, Limonene.

- Anti-psychotic (Tranquilizing) CBD, CBC, THC, CBN, CBG, Myrcene, Linalool.

- Anti-spasmodic- (Suppresses muscle spasms) THCA, THC, CBG, Myrcene.
- Anxiolytic (Relieves anxiety) CBD, CBG, Linalool, Limonene.
- Appetite stimulant (Enhances appetite) THC.
- Neuro-protectant (Retards nervous system degeneration) THC, CBG, CBC, THCA, CBD.
- Bone stimulant (Promotes bone cell growth) CBN, THCV, CBG, CBC.

Remember, these molecules all work together; it is usually best not to just use one.

Chapter 8

Methods of Ingestion of Medicinal Cannabis

With an understanding of the myriads of medicinal molecules in cannabis, methods of ingesting those molecules should be examined. To begin, smoking or combustion is going to be eliminated. No medical professional is going to recommend the combustion of medicine. It's a waste of expensive medicine, and smoking is believed to have health implications. This chapter is designed to provide medical cannabis patients an understanding of the methods of ingestion of cannabis other than inhalation of combusted botanic cannabinoids and terpenes.

Vaporization

The rapid emergence of vaping has changed the landscape of medical cannabis administration, and cannabis vaping devices come in a wide variety of shapes, sizes, and designs. Still, vaporizers can generally be divided into cannabis concentrate vaporizers and dry herb vaporizers. Cannabis concentrates are created through butane extraction and can take several forms, from a thick liquid to a firm, almost glassy solid. These forms are referred to by several slang terms such as 'dabs,' 'budder,' 'earwax,' 'honeycomb,' and 'shatter,' among others (Chadi et al., 2020). Both concentrate vaporizers and dry herb vaporizers vary in terms of size, ranging from small flash-drive-like devices to larger tank-like devices. Mechanisms for both types of devices usually include a power source (often a small battery), a heating element, a vaporization chamber, a cartridge or reservoir, and a mouthpiece. Concentrate vaporizers also contain a cartridge or reservoir, The vaporization process involves heating the compound, producing an aerosol, which is then inhaled and absorbed systemically through the respiratory system (Dinakar & O'Connor, 2016).

Cannabis-concentrate vaping devices share many similarities with e-cigarettes and other nicotine vaping devices. Specifically, cannabis concentrate and e-cigarette devices come with prefilled liquid cartridges that contain nicotine or cannabis concentrates, making it very simple to inhale evaporated molecules of one substance or the other (Fadus et al., 2019). Studies suggest that vaporized cannabis may generate fewer chemicals than smoked cannabis and thus represents a 'healthier' mode of consumption (Loflin & Earleywine, 2015). However, the use of high-potency concentrates, like those found in vape pens, correlates with a higher incidence of mental and physical health problems and may lead to a higher risk of developing acute adverse effects, such as paranoia, psychosis, and cannabis hyperemesis syndrome (Prince & Conner, 2019).

Like most institutions that act as functionaries of the American Healthcare Delivery System, the motivations of dispensaries are based on capitalism. The THC molecule is the cheapest and easiest to extract from the cannabis plant, and because it provides feelings of bliss and euphoria, it is the easiest molecule for dispensaries to push. Analogous to the majority of McDonald's profits being generated by French fries, the majority of dispensary profits are generated by the THC molecule. This is particularly well-illustrated in how they provide topicals, which demonstrate efficacy in treating skin conditions like psoriasis. Despite the cannabidiol (CBD), cannabigerol (CBG), and cannabichromene (CBC) molecules demonstrating greater efficacy for treating skin conditions than THC (Wroński et al., 2023), dispensaries typically only carry THC topicals, and the patient often has to resort to the private market to acquire botanic cannabinoid topicals to treat problematic skin conditions because the appropriate botanic cannabinoid combinations are not available at the institutions established by the American Healthcare Delivery System.

Cannabis concentrates available through the dispensaries established by the American Healthcare Delivery System may exceed 97 percent THC, leaving little room for the other botanic cannabinoids and terpenes. As has been discussed, most patients experience significant pain relief utilizing moderate doses of THC, and high doses have been demonstrated to actually increase pain (Wallace et al, 2007). Furthermore, many concentrate vaping devices use highly processed products whose safety and chemical profile are much closer to that of e-liquids used in e-cigarettes. This is especially true of flavored cannabis vaping products, which may contain several harmful and carcinogenic aerosols (Civiletto & Hutchison, 2023). While there is still much to be discovered pertaining to the short- and long-term effects of cannabis concentrate vaping devices on the lungs, the significant presence of carbonyls, volatile organics, nitrosamines, and heavy metals, all considered toxic and carcinogenic, found in many concentrate vaping devices, provides a reason for concern (Goniewicz et al., 2014; Pankow et al., 2015).

The most intelligent method of ingestion is often the vaporization of dried cannabis flower. There is an erroneous belief among recreational users that vaporization doesn't get you as high as smoking. Most medical cannabis patients are looking for symptom relief and so generally choose strains and doses that don't get them high, but if that is the goal, vaporization is just as effective as smoking. Another belief of recreational users is that you get higher if you breathe the smoke in and hold it in your lungs. This is also not true and could conceivably lead to some health consequences.

Vaporization of dried cannabis flower is the logical choice for most health-conscious cannabis consumers. The problem is that many people have never been taught how to vaporize properly, and the cheaper vaporizers on the market often don't have very good thermostats. Most of the medicinal cannabinoids and terpenes evaporate at around 356 degrees Fahrenheit. Because the medicinal compounds in cannabis are volatile, they can be vaporized at a temperature significantly lower than that needed to reach combustion or smoke. Vaporization exploits this as the patient can draw hot air through the flower, which in turn evaporates the cannabinoids and terpenes and frees them for inhalation. Vaporizers steadily heat the herb to a temperature high enough to extract the THC, CBD, and other botanic cannabinoid and terpene molecules but too low to release the potentially harmful toxins associated with combustion.

Below is a chart showing the evaporation temperature of the various cannabinoids and terpenes but be aware that some of the less expensive vaporizers on the market do not have very reliable thermostats. Two types of dry herb vaporizers exist: desktop and portable. A desktop vaporizer is essentially a box that heats the product to the point that the medicinal molecules evaporate and can be inhaled.

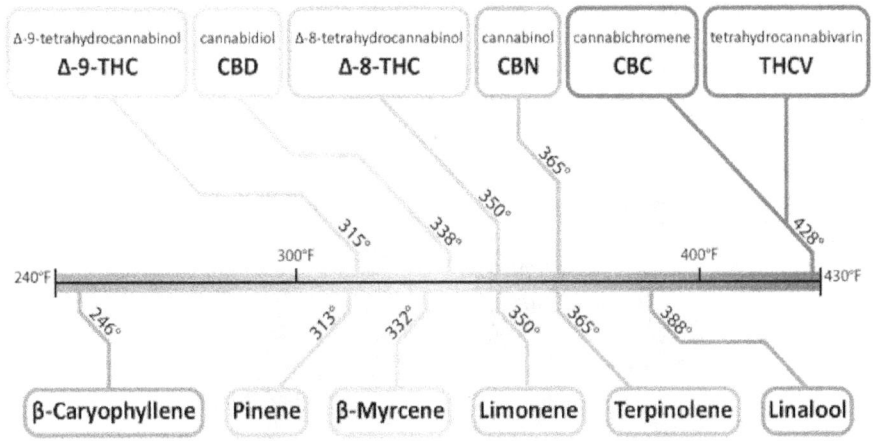

Inexpensive Desktop Dry Herb Vaporizer

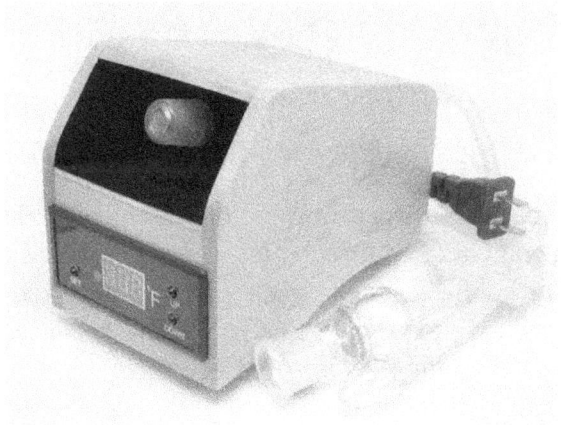

Portable vaporizers are convenient because they are concealable and allow the patient to adapt to the ongoing institutionalized stigmatization of the utilization of natural medicines. The limiting factor of portable dry herb vaporizers is typically battery life.

Portable Dry Herb Vaporizer

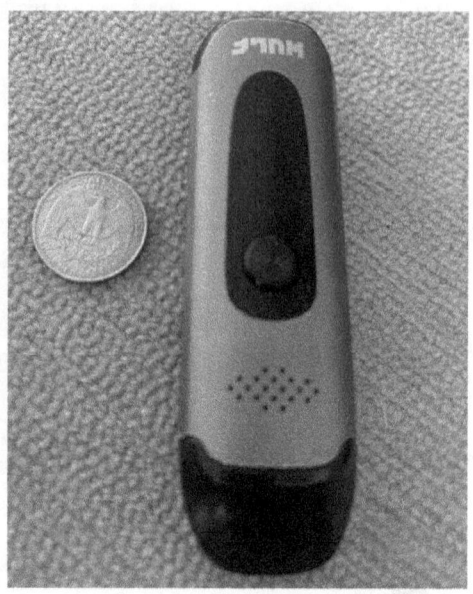

Wulf Next Dry Herb Vaporizer. Available at WulfMods.com

Most patients who use cannabis to treat chronic pain dose with about an eighth of a gram of dried flower. This is about half the size of a quarter and twice as thick. This low dose manages neuropathic and inflammatory pain for a few hours, and patients usually administer as needed to manage their pain issues throughout the day. Mileage may vary, but a moderate dose is important to maintain the balance of the endocannabinoid system.

Studies suggest that vaporized cannabis may generate fewer chemicals than smoked cannabis and thus represents a less harmful or 'healthier' mode of administration of botanic cannabinoid and terpene intromission (Loflin & Earleywine, 2015). Still, the use of high-potency concentrates, like those found in vape pens, correlates with a high incidence of mental and physical health problems and may lead to a higher risk of developing acute adverse effects, such as paranoia, psychosis, and cannabis hyperemesis syndrome (Prince & Conner, 2019). While dry herb vaporizers utilize unprocessed botanic cannabinoids and terpenes, capitalists rather than clinicians have designed and marketed vaping devices that use highly processed products whose safety and chemical profiles are much closer to those of e-liquids used in e-cigarettes. This is especially true of flavored cannabis vaping products, which may contain multiple harmful and carcinogenic aerosols that disperse into the lungs through various combinations of glycerol, propylene glycol, acetaldehyde, and formaldehyde (Civiletto & Hutchison, 2021; Troutt & DiDonato, 2017). The significant presence of carbonyls, volatile organics, nitrosamines, and heavy metals, all considered toxic and carcinogenic, found in cannabis oil vaping products provides reason for concern (Chadi et al., 2020).

Edibles

Most novices come into contact with this method of ingestion, and that can present a problem. The effects of consumed cannabis are much different when compared to vaporizing. Edibles are much slower to kick in and wear off and give more of a full-on "body high," described as "heavier and deeper" than if vaporized. This method of ingestion is particularly beneficial for those with severe body pain. However, with edibles, it's easier to overconsume and therefore overmedicate. The wonderful thing about cannabis is that you can't overdose on it. You'd have to eat 5000 pounds of it to reach a lethal amount, and of course, what would kill you is not the cannabis but the fact that you ate 5000 pounds of something.

The main thing to be aware of in any method of oral consumption of cannabis is the rule, "Start low, and go slow." That means starting

with a small amount and waiting an hour or so to feel the effects before adding any more. Most of the edible cannabis-infused products you find in dispensaries contain 100 milligrams of THC. Often, these are in the form of gummies or candy bars similar to Hershey bars. Hershey bars are divided into sections. These sections are called pips, and in a normal Hershey bar, there are 12. In a cannabis-infused bar, there are usually 10, with 10 milligrams of THC in each one, totaling 100. The threshold for psychoactivity is 5 milligrams, so the simplest strategy for medicating with these is to cut one pip in half and eat it. Cannabis naïve patients should probably cut it in half again and eat that and wait a couple of hours to feel the effects before consuming any more. It may be difficult not to eat a whole candy bar but understand that this is medicine and should be treated as such. Overmedicating is not an enjoyable experience. If you ever do overmedicate, don't panic. Find a quiet place to lie down for a while and think happy thoughts. You'll probably sleep about 8 hours and wake up fine, although possibly somewhat groggy.

It is important to think of cannabis-infused edibles as medication and to treat them that way. Don't leave them lying around for children or pets to find. There are many varieties of cannabis-infused edibles, and they are designed to be appealing to the consumer. Here are some pictures of cannabis-infused products. Around Christmas, a company comes out with cannabis-infused truffles that look really good. Ben and Jerry's plans to come out with cannabis-infused ice cream as soon as it is legal for them to distribute it nationwide.

Dixie One Watermelon Cream
•••••
This refreshing beverage has just 10mg of THC to give you a mild high and...

Old Fashioned Sarsaparilla
•••••
A throwback flavor with a modern kick. Sarsaparilla has been used by healers for ages...

Wild Berry Lemonade
•••••
Cool off with a wild berry twist on the classic lemonade flavor you remember. Dixie's...

Tinctures

Tinctures are a concentrated form of medicinal cannabis, often in an alcohol solution. These are highly concentrated and, again, require careful dosing. They are applied with an eyedropper sub-lingually (under the tongue) and absorbed into the bloodstream that way. Again, this can't be stressed enough. "Start low, go slow." Start out small and wait to feel the effects before adding any more.

Ingestible Oils

This is a Full Extract Cannabis Oil (FECO), often called Rick Simpson oil. Rick Simpson used it to treat his skin cancer. He didn't invent the extraction process but took it to the next level by marketing it. A cultivator in Minnesota prefers the name FECO rather than Rick Simpson oil because, as he put it, "Rick Simpson was a dick to me one time." These ingestible oils can easily be put into capsules, and the pharmaceutical industry has conditioned Americans to take capsules.

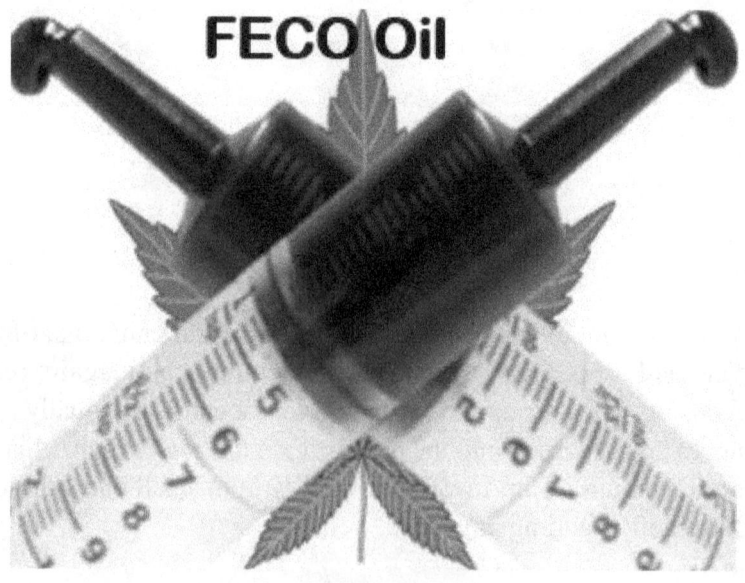

Botanic Cannabinoid Topicals

These are balms, sprays, oils, and creams that are applied to the skin or muscles. These are useful in combating some skin conditions that medical practitioners mostly ignore because they are not life-threatening and difficult to treat effectively. Cannabis topicals are tremendously effective against conditions like psoriasis, joint diseases like rheumatoid arthritis, and everyday stress. Topicals are non-psychoactive, so you could bathe in them and not get high.

Transdermal Botanic Cannabinoid Patches

Compounds have been applied to the skin for thousands of years to enhance beauty and treat local conditions. However, most topically applied compounds and drugs are poorly absorbed or not at all. This is due to the size and polarity of the drug molecule and the skin's barrier effect. Many compounds are meant to remain on the skin surface (topical), such as insect repellents, sunscreens, and antiseptics, while others penetrate the skin layers (transdermal) as a method of entering the bloodstream and avoiding gastrointestinal processes.

Transdermal delivery technology has been developed to treat a wide range of conditions beyond the application site. Transdermal delivery as a mode of intromission offers significant advantages over oral administration due to minimal first-pass metabolism, avoidance of the adverse gastrointestinal environment, and the ability to provide prolonged and quantifiable drug delivery. The FDA has approved transdermal delivery for a wide variety of synthetic medications. Examples include nicotine, lidocaine, colchicine, scopolamine, estradiol and testosterone, nitroglycerine, and fentanyl patches. Despite the advantages, most synthetic drugs cannot be delivered transdermally due to their molecular size and structure (Guy, 2024). Transdermal medications refer to medications that are applied to the skin but involve skin penetration-enhancing compounds or technology that increase the amount of a drug that can cross the skin barrier to the point that the medicine can enter systemic circulation and exert effects in areas other than the site of application (Phatale et al., 2022).

The skin is the largest organ of the human body, comprising roughly 15% of the body's weight (Richardson, 2003). Its primary function is as a barrier between the body and the external environment. This barrier protects against ultraviolet radiation, microorganisms, allergens, chemicals, and water and nutrient loss. Additionally, the skin is involved in many other functions, such as thermal regulation, metabolism, and blood pressure control. The skin also serves as an important sensory organ, providing information about the environment, such as temperature, pressure, and noxious stimulation (pain).

Transdermal drug delivery refers to administering medicinal molecules through the skin into systemic circulation. It is a painless method of delivering medicine systemically. It differs from topical administration in that it delivers medicine across the epidermis, dermis, and subcutaneous tissues and into the systemic circulation at a predetermined and controlled rate (Yewale et al., 2021).

Transdermal delivery in the form of patches provides an FDA-approved administration method for the administration of botanic cannabinoids and terpenes, which the FDA is prohibited from approving due to their non-synthetic origination. Still, this mode of intromission provides an FDA-approved alternative to oral and nonparental drug delivery systems by minimizing and avoiding drawbacks. Oral and nonparental drug delivery systems have limitations, such as a peak and valley phenomenon, resulting in fluctuations in plasma drug levels and causing unpredictable and erratic responses (Alexander et al., 2112). Transdermal patches provide an FDA-approved delivery method for sustained delivery of botanic cannabinoids and terpenes with little risk of side effects while maintaining the drug level above the minimum therapeutic concentration. Furthermore, as a controlled method of drug intromission, transdermal patches are highly convenient, user-friendly and allow for easy termination if necessary (Magnusson et al., 2001).

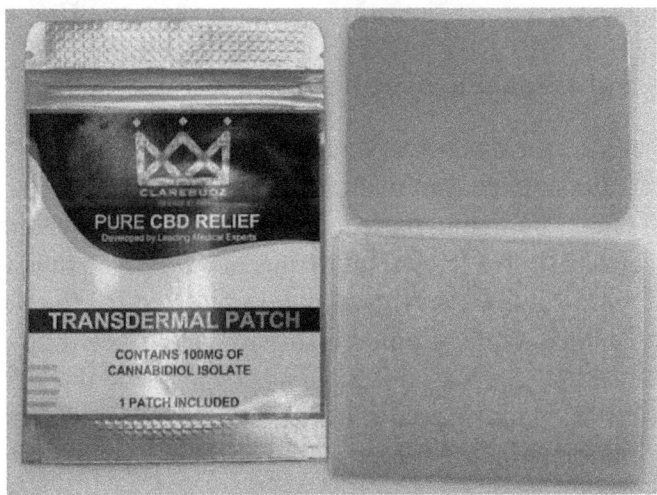

Botanic Cannabinoid Suppositories

A less popular form of cannabinoid intromission is suppositories. There are very few scholarly, peer-reviewed articles that discuss this mode of administration and adhering to the federally mandated synthetic paradigm. Dell and Stein (2021) claim that the synthetic cannabinoid THC-hemisuccinate is the form of THC that has the best rectal absorption. Clinicians believe that suppositories might be helpful for palliative care, for patients who cannot swallow, for gastrointestinal illnesses, and for healing skin damaged by rectal radiation. The onset of action is about 15 minutes, and the effect may last up to 12 hours (Backes, 2017).

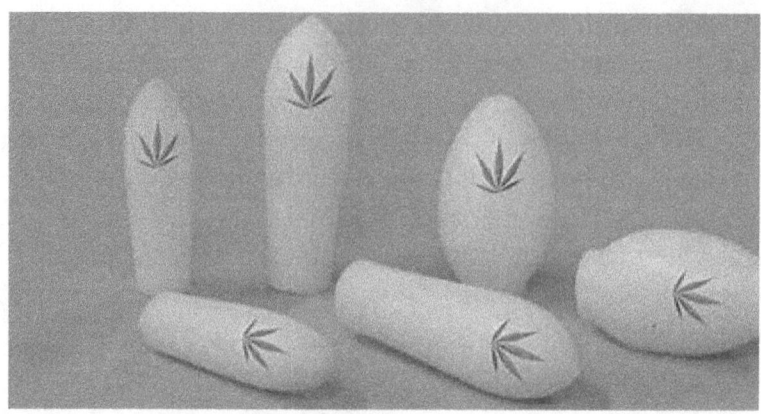

Nanotech and cannabinoid medicine

The increasing global acceptance and interest in the use of botanic cannabinoid and terpene formulations has ushered in a new paradigm of medicine. These medicinal molecules produced by the cannabis plant have been demonstrated to treat numerous diseases. The rapidly shifting landscape and growing body of clinical evidence for the therapeutic efficacy of botanic cannabinoids is due in part to new knowledge on cannabis molecular genetics, decoding its genome, and the identification of CB1 and CB2 cannabinoid receptor crystal structures, along with their heteromeric complex formations. This paradigmatic progress has transformed biological cannabinoid research into mainstream science with remarkable advances through discovering and modulating the human endocannabinoid system (Cristino et al., 2020; Onaivi et al., 2020).

Politicians who make decisions about which aspects of truth scientists are permitted to pursue are beginning to open the door to the research of biologically produced cannabinoids. This blessing of the pursuit of truth by politicians has resulted in scientists beginning to research and understand the complex pharmacology of botanic cannabinoids and the other molecular constituents found in cannabis, with a rapidly advancing understanding of innovative cannabinoid therapeutic options (National Academies of Sciences, Engineering, and Medicine, 2017). Such evidence-based scientific knowledge supports novel approaches for cannabinoid-based nano-formulation and cannabis use in theranostic nanomedicine. The progress in the clinical development of nanomedicines, due in part to advances in nanotechnology, is fueling the development of multifunctional nanocarrier strategies as drug-delivery systems (Mulvihill et al., 2020; Singh et al., 2020). Botanic cannabinoid theranostics are challenging to develop but have potential applications and advantages for several disease conditions that are not adequately addressed with existing pharmaceutical medicines. Theranostics is a two-pronged approach to diagnose and treat cancers using radiotracers. The highly aggressive and invasive glioblastoma tumor is the most malignant lesion among adult-type diffuse gliomas, representing the most common primary brain tumor in the neuro-oncology practice of adults. Embedding micro-particles of THC and CBD for deployment within brain tumor cells allows for the sustained release of botanic cannabinoids at high concentrations directly at the tumor site (Dasram et al., 2024).

Summary

Those are nine methods of ingestion of botanic cannabinoids other than smoking. The advancements in this aspect of medical science are incredible, falsifiable, and arguably the most compelling area of study that a biomolecular psychologist could pursue.

Chapter 9

Diagnosis and Treatment of Endocannabinoid Deficiency Disorders

A Study Proposal

Introduction

On June 18, 1971, as part of his war on drugs, President Richard M. Nixon instituted a ban on the research of biologically produced cannabinoids in the United States (Slaughter, 1988). Two years later, Nixon established the Drug Enforcement Administration (DEA), a paramilitary division of the Department of Justice, and tasked it with ensuring that no research of biologically produced cannabinoid molecules occurred within the confines of the United States of America. When Nixon prohibited pursuits of truth pertaining to biologically produced cannabinoids, the only biologically produced cannabinoids of which humanity was aware were produced botanically by the species *Cannabis sativa*. Devane et al. (1992) discovered the endocannabinoid system two decades later. This system has since been described as the single most important scientific medical discovery since the recognition of sterile surgical techniques (Allen, 2016). As our knowledge expands, it becomes increasingly apparent that the endocannabinoid system is analogous to a master control system of virtually all human physiology (Dawson & Persad, 2022; Pepper et al., 2019; Sallaberry & Astern, 2018).

Statement of the Problem

The problem to be addressed by this study is that the half-century prohibition of research on cannabinoids produced biologically has curtailed studies of cannabinoids produced naturally through human biological processes. Research universities as institutions have resisted the pursuit of truth pertaining to these molecules to avoid jeopardizing their federal funding. This is the nature of science.

On December 2, 2022, President Joe Biden ratified the Medical Marijuana and Research Expansion Act into law, effectively rescinding the Nixon mandate that prohibited any research of non-synthetic cannabinoids in America and further establishing the precedent that the President of the United States has the power to determine which aspects of truth American scientists are permitted to pursue (Purcell et al., 2022).

While this opens the door slightly to allow the research of botanically produced cannabinoids, that door is still heavily guarded by the paramilitary unit known as the DEA. However, there is no longer a guard at the now wide-open door to the pursuit of truth pertaining to the cannabinoids the human body produces naturally. These endogenous lipid-based retrograde neurotransmitters bind to cannabinoid receptors and cannabinoid receptor proteins expressed throughout the vertebrate central and peripheral nervous systems (Devane et al., 1992). To date, basal concentrations of endocannabinoids have never been quantified by researchers in the United States, although techniques for doing so have been repeatedly described by cannabinoid researchers residing in more scientifically progressive countries (Di Filippo et al., 2008; Fanelli et al., 2012; Gachet et al., 2015; Gao et al., 2020; Hill et al., 2008; Krumbholz et al., 2013; Lam et al., 2008; Mwanza et al., 2016; Ney et al., 2023; Ottria et al., 2014; Pertwee, 2014; Romero-Sanchiz et al., 2019; Wilker et al., 2016; Zoerner et al., 2011).

Purpose of the Study

This study aims to quantify basal concentrations of endogenous cannabinoids produced by individuals medically classified as metabolically healthy. Twelve percent of the nation's population is classified as metabolically healthy. Metabolic health is defined as having optimal levels of waist circumference (WC <102/88 cm for men/women), glucose (fasting glucose <100 mg/dL and hemoglobin A1c <5.7%), blood pressure (systolic <120 and diastolic <80 mmHg), triglycerides (<150 mg/dL), and high-density lipoprotein cholesterol (≴40/50 mg/dL for men/women), and not taking any related medication (Araújo et al., 2019).

Assumptions

1. Individuals clinically classified as metabolically healthy have well-functioning and balanced endocannabinoid systems.

2. Delineating basal endocannabinoid ranges of metabolically healthy individuals provides techniques for diagnosing endocannabinoid deficiency disorders.

3. Once diagnosed, endocannabinoid deficiencies can be efficaciously treated through supplementation of phytocannabinoid equivalents.

Materials and Instrumentation

Determining individual basal ranges of currently identified endocannabinoids requires their standards, which are commercially available through a quality scientific chemical supply institution, gas, or liquid chromatography instrumentation in tandem with a mass spectrometer. The technique for quantifying basal endocannabinoid concentrations utilizes a methodology similar to the technique used to determine botanic cannabinoid concentrations produced by cannabis chemovars. Utilizing the necessary endogenous cannabinoid standards and this instrumentation, endocannabinoid tone can easily be quantified by this study on an industrialized scale.

Significance of Study

The determination of basal endocannabinoid ranges of a population of metabolically healthy individuals will provide a technique for diagnosing endocannabinoid deficiency disorders that manifest as autoimmune and mood disorders, such as PTSD, fibromyalgia, Crohn's disease, and others, and can efficaciously be treated through phytocannabinoid equivalent supplementation. The concepts inherent in this study would benefit graduate students in the cooperating research university significantly by providing them with direct experience with mass spectrometry and liquid or gas chromatography instrumentation. Furthermore, a study of this nature will enhance these students' experiences in quantitative research methodology as well as their understanding of the nature, history, and philosophy of science.

Chapter 10

Specific Medical Conditions and Their Treatment with Medicinal Cannabis

Medicinal cannabis can alleviate the symptoms of many ailments but is rarely a cure. However, supplementation of the endocannabinoid system with judicious amounts of plant cannabinoids may reduce the incidence of some diseases and prevent others. The key to the successful use of cannabis is to select the proper dose and frequency. The ailments that follow have been selected because cannabis has been used and shown to be effective for symptomatic relief.

Alzheimer's disease

Alzheimer's disease is an age-related brain disease that is often associated with profound cognitive decline. Key aspects of this disease are tightly intertwined with the body's endocannabinoid system, and soon, measures for the prevention of Alzheimer's will target parts of the endocannabinoid system. Some treatments for this disease will be botanic cannabinoid-based (Backes, 2017). Alzheimer's disease is the leading cause of dementia among the elderly. With the ever-increasing size of the population, cases of Alzheimer's are expected to triple over the next 50 years. Consequently, the development of treatments that slow or halt the disease progression has become imperative to improve the quality of life for patients and reduce the healthcare costs attributable to Alzheimer's disease (Fritze et al., 2019).

In Alzheimer's disease, deposits of a protein plaque called beta-amyloid build up between nerve cells and blocks neurons that are trying to fire messages among themselves. This results in the confusion many Alzheimer's patients seem to suffer. There are some cannabinoids that break this plaque up. One of the most important is the Δ9-Tetrahydrocannabinol molecule (THC), and this is the most critical botanic cannabinoid to target for Alzheimer's patients (Cao et al., 2014).

Botanic Cannabinoids and Terpenes to Target Alzheimer's Disease

When analyzing a strain to target Alzheimer's, the botanic cannabinoids the patient should seek out are THC, CBD, CBC, CBG, and THCA. The flawed paradigm under which the American Healthcare Delivery System operates has resulted in limited efficacy of current therapies against Alzheimer's disease and highlights the necessity for intensifying research efforts devoted to developing novel agents for preventing or retarding the disease process. Targeting the endogenous cannabinoid system has emerged as a potential therapeutic approach in the treatment of treat Alzheimer's. The endocannabinoid system is comprised of cannabinoid receptors, including the well-characterized CB1 and CB2, with their endogenous ligands and the enzymes related to the synthesis and degradation of various endocannabinoid compounds. Results of multiple studies indicate that the activation of both CB1 and CB2 receptors by botanic agonists has beneficial effects on Alzheimer's by reducing harmful β-amyloid peptide action and tau phosphorylation, as well as by activating the brain's intrinsic repair mechanisms. Furthermore, endocannabinoid signaling has been demonstrated to modulate numerous concomitant pathological processes, including neuroinflammation, excitotoxicity, mitochondrial dysfunction, and oxidative stress (Aso & Ferrer, 2014).

Of the terpenes, these molecules are considered cholinesterase inhibitors: Pulegone, alpha-pinene, alpha-terpineol, terpineol-4-ol, and p-cymene (Shereen Lai et al., 2022). Myrcene is also indicated because it synergizes the other terpenes and changes the permeability of the cell membranes to allow better absorption of the cannabinoids into the brain (Jansen et al., 2019). Pulegone is also warranted because it has been shown to have memory-boosting effects. Alpha-Pinene and Eucalyptol are memory-boosting terpenes and should also be targeted (Hashemi & Ahmadi, 2023; Silva-Correa et al., 2021; Surendran et al., 2021). Linalool should also be considered due to its calming effect (Xu et al., 2021). Limonene may also be appropriate due to its potential as a neuroprotectant properties and its tendency to enhance mood (Eddin et al., 2021; Zhang et al., 2019).

Best Method of Ingestion for Alzheimer's Disease

Oral cannabis preparations are considered best because the effects are long-lasting and easily incorporated into a palatable and appealing form for the Alzheimer's patient. Care should be taken not to leave the oral preparations around where they could be ingested by pets or mistaken by the patient as a snack. This is medication and should be treated as such.

Amyotrophic Lateral Sclerosis

Amyotrophic lateral sclerosis, also known as ALS or Lou Gehrig's disease, named after the baseball legend, is a debilitating disorder that affects an estimated 30,000 people in the United States at any given time. ALS is a degenerative neurological disorder that causes muscle weakness, wasting, and paralysis of the limbs, as well as those that control vital functions such as speech, swallowing, and breathing. It is often the deterioration of these crucial muscles that leads to respiratory failure; even with breathing aids or a tracheotomy, the risk of respiratory failure persists (Hardiman et al., 2011).

Being diagnosed with ALS is devastating, and for many, it can feel like a death sentence. The average life expectancy of a person with ALS is two to five years. However, more than half of all people with ALS live more than three years from diagnosis, and there are outliers, such as the brilliant theoretical physicist Stephen Hawking, one of the most iconic figures with ALS. Hawking was diagnosed in 1963 when he was a 21-year-old graduate student at Cambridge University. He died in 2018 at the age of 55 and is among only 5% of those diagnosed with ALS who lived more than 20 years after being diagnosed.

Cannabis has long been known as a viable treatment option to relieve symptoms of ALS, but the most astonishing results have come from several ALS sufferers who have managed to slow the progression of the disease through regular, controlled doses of cannabis. A remarkable case comes from Cathy Jordan, who was diagnosed with ALS in 1986 and given less than five years to live. In the winter of 1989, Jordan spent the holiday in Florida, preparing for the end of her life, when she made a crucial discovery. While walking on the beach one night, she smoked a joint of Myakka Gold. She felt her symptoms cease, essentially experiencing the neuroprotective effects of botanic cannabinoids before they'd been proven.

Jordan never set out to be a cannabis activist, preferring instead to continue treating her disease quietly. She tried to tell her neurologist in 1989 that cannabis had helped her and adhering to the flawed paradigm of the American Healthcare Delivery System, he tried to convince her husband to have her committed to a mental facility. In 1994, Jordan met a new doctor who was astonished by her progress. When he asked her what she had been doing to stay alive, she informed him, and he advised her, "Smoke all the cannabis you can and never tell a soul because they will never believe you." Unfortunately, Donnie Clark, the grower of Myakka Gold (named for the Myakka region of Florida), was arrested and sent to prison for 12 years until his sentence was commuted by former President Clinton. The strain itself, which Jordan credited with stopping the progression of her ALS, has since been eradicated by the DEA.

Cathy Jordan, however, ended up becoming the inspiration for Amendment 2, the medical cannabis initiative in Florida that was narrowly defeated in the November 2016 election. Ironically, on February 25, 2013, the same day that the Cathy Jordan Medical Cannabis Amendment was announced, Jordan's home was raided. Jordan was arrested, and 23 plants were seized by local authorities, although charges were dropped when it became evident that she was using cannabis for medicinal purposes.

Finally, years later, there has been more and more conclusive research indicating that these patients' self-medicating regimen was efficacious after all. Multiple studies conclusively demonstrate that botanic cannabinoids have powerful anti-oxidative, anti-inflammatory, and neuroprotective effects. Regular applications slow the progress of the disease and prolong the lives of the individuals affected by ALS (Atalay et al., 2019; Cassano et al., 2020; Viana et al., 2022). Amyotrophic Lateral Sclerosis is a qualifying condition for virtually every state with a medical cannabis program. The decision to include it as a qualifying condition is based on the politician the state's citizens elect to make their medical decisions.

Botanic Cannabinoids and Terpenes to Target Amyotrophic Lateral Sclerosis (ALS)

Cannabigerol (CBG) must obviously be included in this discussion because it synergizes the other cannabinoids (Russo, 2011). It has also been found to stimulate brain and bone cell growth (Khajuria et al., 2023; Nachnani et al., 2021). ALS is a painful condition, and so these cannabinoids are indicated: Δ9-THC, CBD, CBC, and THCV. CBC, Δ9-THC, and CBG. All are anti-depressants, and THCA has antispasmodic properties. Δ9-tetrahydrocannabinol (Δ9-THC), cannabinol (CBN), cannabigerolic acid (CBGA), cannabinolic acid (CBDA), and Δ9-tetrahydrocannabinolic acid (Δ9-THCA) should be included in this mix because they are all potent anti-oxidant agents (Dawidowicz et al., 2021). Although the findings are mixed, these botanic cannabinoids also appear to demonstrate modest antispasmodic activity (Nielsen et al., 2019).

The terpenes to target pain are borneol, pinene, myrcene, and eucalyptol for their antinociceptive properties (Schwarz et al., 2023). For depression, linalool and limonene should be included for obvious reasons. α-terpineol is the terpene that comes to mind as an antioxidant and should also be targeted. It is obviously impossible to locate a chemovar of cannabis that has a high percentage of all these molecules, but this provides a basic idea of the botanic molecules the ALS patient should target.

Best Method of Ingestion for Amyotrophic Lateral Sclerosis

Vaporization is probably best because the medicinal benefits are felt immediately. Edibles should also be considered due to their long-lasting effects, and as usual, moderate dosing is essential.

Anxiety Disorders

In humanity's war against mood disorders such as anxiety, botanic cannabinoids and terpenes have been utilized for thousands of years. It is essential that moderate doses of intoxicating botanic cannabinoids such as THC be used because there is considerable evidence that large doses can increase dopamine levels to the point of triggering anxiety and paranoia in susceptible individuals.

The first cannabinoid that should be considered when it comes to treating anxiety is cannabidiol (CBD). CBD kills anxiety like Raid kills cockroaches (Backes, 2017). Additionally, CBG and CBC should be targeted because they inhibit the uptake of GABA, resulting in a decrease in anxiety (Clarke et al., 2021). A tiny amount of THC should likely be included in the mix because of its ability to mitigate the purported depressive effects of the other two cannabinoids.

Terpenes are being used worldwide as a potential source of novel drugs for anxiety disorders. Some of the terpenes in cannabis that have anti-anxiety properties are linalool and myrcene (Kahn et al., 2022; Weston-Green et al., 2021). Limonene should be targeted for its mood-enhancing qualities, and linalool for its calming effect.

Best Method of Ingestion for Anxiety Disorders

Vaporization is probably the best method of ingestion for anxiety because the patient can quickly learn to titrate the proper dose.

Arthritis

Arthritis is one of the earliest ailments for which cannabis was used as treatment. Arthritis covers a wide range of inflammatory conditions but typically refers to two forms of joint inflammation. Rheumatoid arthritis (RA) is an autoimmune disorder characterized by serious inflammation of a joint's interior lining. RA can cause chronic severe pain, permanent joint damage, and disability. Osteoarthritis of the bones is characterized by loss of cartilage in the joints, typically the hands, hips, knees, and spine. Common osteoarthritis symptoms include pain, stiffness, loss of motion, and deformation of the joints. Botanic cannabinoids are effective in treating pain for most patients but may not be well tolerated by older, more cannabis-naive arthritis patients.

The ability of botanic cannabinoid medicines to distract from arthritis pain is well established. Various botanic cannabinoids elicit a range of anti-inflammatory responses. It has been established that the endocannabinoid system and its receptors are found in the synovial membrane of joints. It is believed that botanic cannabinoids play a role in the protection of cartilage in the joints (Richardson et al., 2008).

Botanic Cannabinoids and Terpenes to Target Arthritis

THC is the main pain-relieving botanic cannabinoid. In addition, THC has been cited repeatedly as having twice as much anti-inflammatory activity as hydrocortisone (French et al., 2017; Zurier & Burstein, 2003). Patients suffering from both forms of arthritis would benefit by targeting this botanic cannabinoid; again, a moderate dose is the key. The powerful anti-inflammatory effects of THC and CBD are useful in controlling pro-inflammatory factors secreted by cells associated with tissue damage that occurs in several forms of arthritis. When you really get down to it, it is difficult to find a medicinal compound in cannabis that doesn't have anti-inflammatory properties. This includes many of the terpenes as well as these minor cannabinoids: THCA, CBN, CBDA, CBC, CBN, and CBG. CBG synergizes the anti-inflammatory effects of the other cannabinoids and has anti-inflammatory properties of its own, so it should be targeted along with the others when addressing the symptoms associated with arthritis. Of course, these compounds also have pain-killing aspects, so cannabis ingested medicinally at proper doses is particularly effective against most forms of arthritis. Dosage of THC for arthritis pain should follow the "sweet spot" model for cannabis-induced distraction from pain. Start with 5 milligrams of THC and slowly increase the size of subsequent doses until pain relief peaks.

Regarding essential oils (terpenes), the anti-inflammatories are beta-myrcene, beta-caryophyllene, Alpha-pinene, borneol, eucalyptol, and turpentine. These terpenes should be targeted. All these terpenoids are included for their pain-relieving and anti-inflammatory properties. Eucalyptol has some powerful antinociceptive properties as well and should also be targeted.

Best Method of Ingestion for Arthritis

Different ingestion techniques should be employed for rheumatoid as opposed to osteoarthritis. Oral cannabis preparations are an excellent choice for rheumatoid arthritis because their effects are long-lasting. For osteoarthritis, vaporized cannabis is particularly effective because the medicinal benefits are felt immediately.

Asthma

It may seem counterintuitive that cannabis has properties to treat asthmatic symptoms, but this has been demonstrated in studies dating back to the 1970s (Rosenkrantz & Fleischman, 1978; Tashkin et al., 1975). Asthma is an inflammatory condition of the airways characterized by bronchospasm and airflow obstruction. Both genetic predisposition and environmental factors play a role in asthma. As cited in a 2000 study in Nature, THC exerts a strong bronchodilation effect when vaporized. As noted in the Nature study, both airway dilation and spasm response are controlled by the endocannabinoid system.

In 1975, Donald Taskin et al. of the University of California Los Angeles published the results of a well-designed study on otherwise healthy patients with stable bronchial asthma. Taskin had these participants exercise until they suffered an acute asthma bronchospasm. During the attack, the patients inhaled either placebo cannabis or 2% THC cannabis. The group receiving the placebo took 30 to 60 minutes to recover from the bronchospasm while the group receiving the actual cannabis recovered immediately. Taskin et al. (1975) effectively treated bronchospasm with cannabis containing only 2% THC, only 1/8th as potent as today's average medicinal cannabis. The Taskin study supports the idea that very little THC is required to dilate the airways. Later studies put the optimal dose at only 200 micrograms. This is well below the threshold for psychoactivity (Williams et al., 1976).

Botanic cannabinoids and Terpenes to Target Asthma

THC-dominant varieties high in pinene are suggested because pinene is a bronchodilator (Kalová et al., 2016; Rohr et al., 2002).

Best Method of Ingestion for Asthma

Tinctures of cannabis were a popular asthma treatment in the 19th century. However, if swallowed, they will take too long to work to be effective for an acute asthma attack because of the length of time it takes to metabolize a swallowed dose. A transdermal patch or sublingual (under the tongue) dose is likely more effective because they enter the bloodstream quickly without having to be processed through the liver.

Some patients react to cannabis vapor with bronchospasms, so care must be exercised. Start with a very small inhalation when stable, before bronchospasm, to gauge how it might be tolerated. Additionally, it is important to use extremely clean cannabis, low in microbial and mold/yeast counts, since these pathogens can irritate the airways or cause secondary lung infections. This is yet another reason that laboratory analysis is important.

Attention Deficit Hyperactivity Disorder (ADHD)

Attention deficit hyperactivity disorder (ADHD) is a term used for a group of behavioral symptoms, which include inattentiveness, hyperactivity, and impulsiveness. ADHD is not usually diagnosed until six to twelve years of age. It affects around 5% of children and 3% of adults worldwide (Ahn et al., 2008). The disorder is characterized by developmentally inappropriate and impairing levels of inattention, hyperactivity, and impulsivity, commonly accompanied by emotional dysregulation, cognitive impairments, and psychiatric comorbidities (Cooper et al., 2017). Behavior genetics, molecular genetic studies, and biomolecular psychology have converged to demonstrate that both genetic and nongenetic factors contribute to the development of ADHD (Asherson et al., 2016). Family, twin, and adoption studies compellingly demonstrate that genes significantly mediate susceptibility to ADHD. These gene studies of ADHD have produced substantial evidence implicating several genes in the etiology of the disorder.

ADHD as an Endocannabinoid Deficiency Disorder

Human and animal studies have consistently demonstrated that the endocannabinoid system is fundamental for emotional homeostasis and cognitive function (Ibarra-Lecue et al., 2018). All vertebrates possess a measurable endocannabinoid tone reflecting concentrations of anandamide (AEA) and 2-arachidonoylglycerol (2-AG), which have been categorized as centrally acting endocannabinoids, and their decreased concentration shows a significant correlation to the development of a variety of physical and psychological disorders (Faraone et al., 2005). Deficiencies of different endocannabinoid system elements contribute to the pathophysiology of several mental disorders, with varying alterations in gene and protein expression of CB1 receptors being demonstrated, depending on the technical approach used or the brain region studied ((Ibarra-Lecue et al., 2018).

The endocannabinoid system has been implicated in various dopamine-deficiency-related disorders, including ADHD (Centonze et al., 2009), autism (Karhson et al., 2018; Su et al., 2021), schizophrenia (Su et all., 2021), Parkinson's disease (Stampanoni Bassi et al., 2017), and Huntington's disease (Laprairie et al., 2015). A complex interaction between the endocannabinoid system and dopamine production has been demonstrated experimentally, displaying the dysfunction of the dopamine system as a proposed explanation of the clinical manifestations of ADHD (Giuffrida et al., 1999).

Pharmaceutical Approaches to Treating ADHD

Typically, the first course of treatment for ADHD is a prescription for an amphetamine (Adderall) or methylphenidate (Ritalin). Dexedrine is twice as potent as Ritalin, a better-known stimulant, and second only to Ritalin in treating ADHD. However, because the Physician's Desk Reference (PDR) lists Dexedrine under "Diet Control" medications, many insurance companies will not cover Dexedrine to treat ADHD (Antshel et al., 2011). While all these synthetic options are addictive and produce a host of unpleasant side effects involving the cardiovascular and central nervous systems, gastrointestinal problems, pituitary dysfunction, blood pressure problems, anxiety, headaches, delayed growth, and nausea, they are considered viable treatment options because they activate the release of dopamine, the key chemical in the brain's reward center (Shier et al., 2013). Dopamine is commonly linked with the brain's pleasure system, providing a feeling of enjoyment and motivation to perform specific tasks (Wilker et al., 2016). ADHD patients have been confirmed to lack appropriate dopamine levels, which allows for the experience of a sense of reward and motivation (Blum et al., 2008). An overwhelming body of neurochemical evidence unequivocally demonstrates that specific phytocannabinoids increase anandamide concentrations, the endocannabinoids responsible for releasing dopamine (Chen et al., 1993; Tanda et al., 1997; Ton et al., 1988; Volkow et al., 2009). Since the discovery of the endocannabinoid system in 1992, extensive anecdotal evidence, survey data, research studies validate that increasing the level of the endocannabinoid anandamide can directly help with not only the side effects of Adderall but also symptoms of ADHD through the direct release of dopamine (Beltramo et al., 2000; Volkow et al., 2009).

Second Line of Pharmaceutical Medications

While stimulants are the first-line medication in the psychopharmacological treatment of attention-deficit hyperactivity disorder, 10% to 30% of all children and adults with ADHD either do not respond well, do not tolerate treatment with stimulants, or demonstrate adverse side effects that are pharmacologically treated with antidepressants such as Zoloft, Paxil, and Prozac (Capuano et al., 2014). Atomoxetine was the first non-stimulant approved for treating ADHD in the United States by the Food and Drug Administration and is marketed as a good alternative for people who cannot tolerate or do not experience the desired effects of stimulant medications. However, there are critical safety concerns about this ADHD drug being implicated in sudden cardiac death and suicidal behavior (Capuano et al., 2014).

Competing Approaches for Treating ADHD

Pharmaceutical and nutraceutical approaches for enhancing dopamine levels to treat ADHD compete in remarkable ways. Pharmaceuticals have the disadvantage of iatrogenic effects, often resulting from the body's inability to degrade the synthetic molecules of which they are comprised. The psychotropic effects produced by synthetic amphetamines have the advantage of claiming FDA approval. Nutraceuticals have the advantage of being natural, providing them some biomolecular superiority. Because they are natural, nutraceuticals have the disadvantage of being unpatentable and cannot be considered for FDA approval (Santini et al., 2018). Since the FDA's inception, the public has been conditioned to accept that FDA-approved medications are safe; however, this perception is being questioned as the adverse effects of FDA-approved treatments are increasingly being exposed.

Shortly after the endocannabinoid system was discovered, Fatty acid amide hydrolase (FAAH) was demonstrated to be the enzyme that degrades anandamide, the dopamine-releasing endocannabinoid (Oz et al., 2010). At that time, molecular engineers began studying FAAH as a target for pharmaceuticals because controlling FAAH levels may yield some of the same health effects that excite clinicians about the potential for phytocannabinoid-based medicines. Synthetic cannabinoids work by inundating the system with molecules structurally similar to THC and other phytocannabinoids (Deutsch & Chin, 1993). Medicines that inhibit the body's manufacture of FAAH are confirmed to have a comparable effect by exploiting the level of deficient endocannabinoids in the central nervous system. If the deficiency is in anandamide, reduced FAAH results in increased concentrations of anandamide (Dawson, 2019b).

Adverse Effects of Synthetic FAAH Inhibitors

Raising the endocannabinoid concentrations by impeding FAAH and additional catabolic enzymes, rather than controlling exogenous agents, is conjectured to reduce cannabinoid-like adverse events attributed to the intromission of a particular phytocannabinoid (Dawson, 2018). Synthetic FAAH inhibitors display neurological side effects not manifested by the biologic, including impairment of cognition and motor functions and a predisposition to psychoses, especially when these agents are used for long-term treatment (Ahn et al., 2008; Martín-Sánchez et al., 2009).

The creation of potent and safe synthetic FAAH inhibitors has been hindered by their harmful side effects (Butler & Callaway, 2016). On July 9, 2015, the Biotrial Research organization initiated human test trials of the synthetic FAAH inhibitor BIA 10-2474 by recruiting 128 healthy volunteers, consisting of men and women aged 18 to 55 (Kaur et al., 2016). The study involved a three-stage design, with 90 participants intromitting the medicine throughout the trial's initial stages. No serious adverse events were reported. The study participants stayed at Biotrial's treatment center for two weeks, undergoing tests, and the pharmaceutical was administered for ten days. Starting on January 7, 2016, six male volunteers received doses by mouth in the third stage. On January 10, the first volunteer was hospitalized, became brain dead, and died a week later. The other five volunteers were also hospitalized. Four suffered injuries, including severe necrotic and hemorrhagic lesions, as displayed on brain MRIs. The trial was discontinued on January 11, 2016. Three of the four men displayed neurological indicators detrimental enough to paint a picture to fear that there would be an irreversible handicap even in the best scenario (Kaur et al., 2016).

Magnetic-resonance-imaging scans revealed bleeding and dying tissue deep within the brain (Kaur et al., 2016). Many questions remain unanswered, particularly about the biomolecular mechanism causing the participants' injuries. This clinical trial's devastating result led to a scramble of scientists suggesting various accounts as to the origin of the deadly adverse effects of the synthetic FAAH inhibitor. It has been proposed that the adverse effects may come from its binding to unknown off-targets. However, few methods exist to foresee cellular off-target effects resulting from the drug binding to biological assemblies and their associations with diseases (Dider et al., 2016). Due to the lack of understanding of the endocannabinoid system by scientists barred from learning about it in college, they missed the mark when attempting to explain what the off-targets of FAAH inhibitors might be and how these off-targets affected the system-level response (Mallet, et al., 2016).

Degradation of Synthetic and Biologic FAAH inhibitors

FAAH inhibitors are designed to remove FAAH proportionally, thereby increasing the concentration of anandamide naturally produced by the body. Enough is now known about biomolecular psychology and the endocannabinoid system to hypothesize about the mechanism by which synthetic molecules cause neurological damage (Dawson, 2019b). The side effects are likely not a byproduct of FAAH inhibition directly but rather the result of biological enzymes being incapable of effectively degrading synthetic FAAH inhibitors. Biologic FAAH inhibitors demonstrate significant variances in their molecular composition compared with their synthetic counterparts (Dawson, 2018). The differences in the molecular structures may explain variances in the safety profiles between the artificial and the biological. These differences are related to the time it takes for the FAAH inhibitors to degrade. Information is lacking about what enzyme degrades synthetic or biologic FAAH inhibitors, and this is an area where further research is warranted.

Technological restrictions, coupled with a suppression of research on biologically produced cannabinoids at many major research universities, have resulted in an inadequate understanding of the endocannabinoid system. A difference in these degradation rates would clarify the variance in adverse effects between biological and synthetic FAAH inhibitors. Despite robust and well-accepted evidence regarding the efficacy of supplementing botanic cannabinoids to treat deficiencies of endocannabinoids, the application of this knowledge is still in its beginning stages (Hill et al., 2018; Morena et al., 2016; Ney et al., 2019).

Comparing Pharmaceutical and Nutraceutical Approaches to Treating ADHD

What we know about manipulating endocannabinoid tone in the treatment of ADHD is elementary but somewhat convoluted. The endocannabinoid primarily responsible for the release of dopamine is anandamide, and increased levels of this molecule demonstrate therapeutic value (Zou & Kumar, 2018). While dopamine deficiency has long been known to be a significant contributor to ADHD manifestation, a chronic deficit of serotonin has also been shown to trigger symptoms of ADHD (Banerjee & Nandagopal, 2015).

The synthetic approach provides two methods of increasing dopamine levels in subjects with ADHD: one exogenous (stimulants and amphetamines) and one endogenous (FAAH inhibitors). Both approaches result in undesirable side effects resulting from the body's inability to degrade them. The nutraceutical approach allows for both exogenous and endogenous techniques for increasing dopamine, and the organic composition of biological molecules allows the body to efficiently degrade them, eliminating these adverse effects. Synthetic stimulants and amphetamines are thought to work because they increase dopamine levels exogenously, and synthetic FAAH inhibitors work by inhibiting anandamide's degradation in the nervous system, endogenously increasing the production of dopamine (Dawson, 2018).

Nutraceutical methodology increases the production of dopamine in similar ways. The phytocannabinoids CBC, CBN, THC, and CBG activate CB1 receptors, thereby exogenously increasing dopamine concentrations, while CBD exogenously releases serotonin (de Mello Schier et al., 2014). Natural FAAH inhibitors work endogenously by inhibiting the degradation of anandamide in the nervous system, thereby increasing dopamine production.

Summary

Research has consistently demonstrated that the endocannabinoid system is fundamental for emotional homeostasis and cognitive functions correlating to the development of various physical and psychological disorders (Ibarra-Lecue et al., 2018). Behavior genetics, molecular genetic studies, and biomolecular psychology have converged to reveal that both genetic and nongenetic factors contribute to the development of ADHD. Targeting the endocannabinoid system in treating ADHD exposes the dopamine system's dysfunction as a proposed explanation of the clinical manifestations of ADHD (Giuffrida et al., 1999).

Current pharmaceutical options that treat ADHD, such as Ritalin and Adderall, are addictive and produce a host of unpleasant adverse events;

however, they are considered viable treatment options because they activate the release of dopamine (Giuffrida et al., 1999). An astonishing 10% to 30% of all children and adults with ADHD either do not respond well or do not tolerate treatment with stimulants or display adverse effects. These side effects are treated with antidepressants, with Zoloft, Paxil, and Prozac being the most widely prescribed (Shier et al., 2013).

Research studies validate that increasing the endocannabinoid anandamide (AEA) level can directly combat the side effects of Adderall and symptoms of ADHD by the direct release of dopamine (Beltramo et al., 2000). Ample neurochemical evidence unequivocally establishes that specific phytocannabinoids increase anandamide concentrations, the endocannabinoids responsible for releasing dopamine (Chen et al., 1993; Di Marzo et al., 1998; Malone & Taylor, 1999; Tanda et al., 1997; Ton et al., 1988; Volkow et al., 2009). ADHD patients have been confirmed to lack appropriate dopamine levels, which allows for the experience of a sense of reward and motivation (Beltramo et al., 2000).

Nutraceuticals have the disadvantage of being unpatentable because they are natural and cannot be considered for FDA approval. The psychotropic effects produced by synthetic amphetamines have the advantage of claiming FDA approval. The public has been conditioned to believe "FDA approved" means safe, although this perception is becoming questioned as adverse effects of FDA-approved medications are increasingly exposed (Santini et al., 2018).

Fatty acid amide hydrolase (FAAH) is identified as the enzyme that degrades anandamide, the dopamine-releasing endocannabinoid (Di Marzo & Maccarrone, 2008). Medications that inhibit the body's production of FAAH are hypothesized to have a similar effect by increasing the concentration of deficient endocannabinoids in the central nervous system. If the deficiency is in anandamide, reduced FAAH results in its increased concentration (Dawson, 2019b).

Synthetic FAAH inhibitors exhibit neurological adverse events not manifested by the biologic, including impairment of motor functions, cognition, and a predisposition to psychoses, especially when these agents are used for long-term treatment (Ahn et al., 2008; Martín-Sánchez et al., 2009). The creation of potent and safe synthetic FAAH inhibitors has been hindered by their harmful side effects (Butler & Callaway, 2016). Scientists missed the mark when attempting to explain the off-targets of FAAH inhibitors and how these off-targets affected the system-level response (Mallet et al., 2016).

Enough is now known about the endocannabinoid system and biomolecular psychology to theorize about the mechanism by which synthetic compounds cause neurological damage (Dawson, 2018). Information is lacking concerning what enzyme degrades either synthetic or biological FAAH inhibitors. This is an area where further research is warranted. A difference in these degradation rates would clarify the variances in the synthetic and biological FAAH inhibitor's adverse effects. The science regarding the efficacy of supplementing botanic cannabinoids to treat endocannabinoid deficiency disorders is robust and well accepted; however, utilization of this knowledge is still in its infancy (Hill et al., 2018; Morena et al., 2016; Ney et al., 2019). The synthetic approach provides two methods of increasing dopamine levels in subjects with ADHD: one exogenous (stimulants and amphetamines) and one endogenous (FAAH inhibitors). Both approaches result in undesirable side effects resulting from the body's inability to degrade them. A chronic deficit of serotonin has also been shown to trigger symptoms of ADHD (Banerjee & Nandagopal, 2015). The nutraceutical methodology increases the production of dopamine in ways similar to the pharmaceutical approach. The phytocannabinoids CBC, CBN, THC, and CBG activate anandamide receptors, thereby exogenously increasing dopamine concentrations, while CBD exogenously releases serotonin (de Mello Schier et al., 2014).

Botanic Cannabinoids and Terpenes to Target ADHD

The phytocannabinoids CBC, CBN, THC, and CBG activate anandamide receptors, thereby exogenously increasing dopamine concentrations, while CBD exogenously releases serotonin. Intromission of these botanic cannabinoids alleviates the symptoms of ADHD nutraceutically by releasing dopamine, utilizing similar mechanisms to those employed by the synthetic drugs produced by the pharmaceutical industry, sans the iatrogenic effects that accompany all medicines produced synthetically. Regarding terpenes, research highlights terpenes like linalool, limonene, myrcene, β-caryophyllene, and α-pinene for their anxiolytic and sedative properties, which may help reduce ADHD symptoms like anxiety and hyperactivity (Hergenrather et al., 2020).

Best Method of Ingestion

A variety of pharmaceuticals are available in transdermal patch form, and this FDA-approved delivery method can readily be appropriated to boost the bioavailability of nutraceuticals (Patel & Shah, 2018). Typically, ADHD patients utilize tinctures as a delivery method due to the rapid uptake of cannabinoids into the bloodstream from beneath the

Tongue. The effects of swallowed oral cannabis tend to be too sedating. Vaporization of dried flower may be beneficial because the immediacy of the effects allows the patient to learn to titrate the most appropriate dose for their body chemistry.

Autism Spectrum Disorders

Autism spectrum disorder is characterized by persistent deficits in social communication and social interaction in multiple contexts, associated with the presence of restricted and repetitive patterns of behavior, interests, or activities (American Psychiatric Association, 2013). Autism is not a disease but a syndrome with multiple genetic and nongenetic causes (Muhle et al., 2004). Children with autism commonly exhibit hyperactivity, self-harm, aggression, restlessness, anxiety, and sleep disorders (Mannion & Leader, 2013).

Conventional pharmaceutical treatments consist of multiple synthetic psychotropic drugs, including atypical antipsychotics, selective serotonin reuptake inhibitors, stimulants, and anxiolytics. These do not treat autism but aim to eliminate inappropriate behaviors, including psychomotor agitation, aggressiveness, and symptoms of obsessive-compulsive disorders (Canitano & Scandurra, 2008; Hurwitz et al., 2012; Stachnik & Gabay,2010; Wink et al., 2010). These synthetic medicines cause severe iatrogenic effects, including nephropathy, hepatopathy, and metabolic syndromes, among others (Silva et al., 2022). This carries a high cost for society and the individual, entailing the reduction of life expectancy by 20 years in patients with autism compared to the population average (Hirvikoski et al., 2016).

In addition to pharmacological treatments, researchers have begun exploring complementary alternative therapeutic approaches, including the use of botanic cannabinoids and terpenes derived from Cannabis sativa (Silva et al., 2022). Botanic cannabinoids are being used to treat symptomatic conditions of autism (Aran et al., 2019; Barchel et al., 2019). CBD and other compounds in the cannabis plant interact with the endocannabinoid system and modulate different aspects related to cognition, socioemotional responses, susceptibility to seizures, nociception, and neuronal plasticity, which are often altered in autism (McLaughlin & Gobbi, 2012; Marco & Laviola, 2012; Marsicano & Lutz, 2006; Trezza et al., 2012).

Botanic cannabinoids and terpenes exhibit demonstrated effects on social behavior in humans (Aran et al., 2019). They enhance interpersonal communication (Salzman et al., 1977) and decrease hostile feelings (Salzman et al., 1976). Δ9-tetrahydrocannabinol (THC) and cannabidiol (CBD) are currently considered to be the principal phytocannabinoids in cannabis due to their influences on the CB1 receptors, which are highly expressed in the frontal cortex and subcortical areas associated with social functioning (Glass et al., 1997). The CB1 receptor and its endogenous ligands, anandamide and 2-AG, regulate social play and anxiety in animal models (Moreira et al., 2008; Seillier et al., 2013; Trezza et al., 2012) and in humans (Gunduz-Cinar et al., 2013; Phan et al., 2008). Genetic and Pharmacological experiments in animal models indicate that social reward is regulated by oxytocin-dependent activation of the endocannabinoid system in the nucleus accumbens, the primary reward center in the brain (Wei et al., 2015). Furthermore, the endocannabinoid system is a critical excitatory and inhibitory modulator of synaptic plasticity through long-term potentiation and long-term depression (Lee et al., 2015).

The endocannabinoids anandamide and 2-AG act as retrograde signaling messengers. Both endogenous ligands reduce the release of neurotransmitters into the synaptic cleft and mitigate overactive brain circuits (Lee et al., 2015). Reduced endocannabinoid "tone" has been demonstrated to be involved in the pathogenesis of autism spectrum disorder in multiple animal models (Busquets-Garcia et al., 2013; Foldy et al., 2013; Jung et al., 2012; Kerr et al., 2013; Melancia et al., 2018; Wei et al., 2016). Activation of the endocannabinoid system in these and other models of autism spectrum disorder reversed the autistic symptoms (Busquets-Garcia et al., 2013; Doenni et al., 2016; Gomis-González et al., 2016; Gururajan et al., 2012; Kaplan et al., 2017; Kerr et al., 2016; Melancia et al., 2018; Servadio et al., 2016; Wei et al., 2016).

Botanic Cannabinoids and Terpenes to Target Autism Spectrum Disorders

THC and CBD have demonstrated efficacy in treating autism spectrum disorders, the former due to its usefulness in enhancing anandamide concentrations and the latter due to its effectiveness in alleviating anxiety. Myrcene and linalool are appropriate terpenes to target symptoms of autism due to their anti-anxiety and neuroprotective effects.

Best Methods of Ingestion

Typical cannabis ingestion methods cannot be recommended for children and adolescents due to uncertainties related to dosing, and most would upset child welfare

authorities to an extraordinary degree. While the FDA cannot approve naturally produced molecules as medicine because they cannot be patented, they have approved delivery methods for botanic cannabinoids and terpenes in the form of transdermal patches. Transdermal patches utilize the same delivery mechanism as FDA-approved nicotine patches while providing for dose control.

Autoimmune Disorders

The endocannabinoid system plays a key role in many autoimmune disorders. Cannabinoids have been demonstrated to have various effects on bodily systems. Through CB1 and CB2 receptors, amongst others, they modulate neurotransmitter and cytokine release. Additionally, current research on the role of botanic cannabinoids in the immune system demonstrates that they possess immunosuppressive properties. They can inhibit the proliferation of leucocytes, induce apoptosis of T cells and macrophages, and reduce secretion of

pro-inflammatory cytokines. In mice models, they are effective in reducing inflammation in arthritis and multiple sclerosis and have a positive effect on neuropathic pain and type 1 diabetes mellitus. Botanic cannabinoids are an efficacious treatment for fibromyalgia and have an anti-fibrotic effect in scleroderma. These findings indicate that botanic cannabinoids are promising immunosuppressive and anti-fibrotic agents in the treatment of autoimmune disorders. Still, studies in human models are scarce and inconclusive, and more research is required in this field (Katchan et al., 2016).

Botanic Cannabinoids and Terpenes to Target Autoimmune Disorders

For symptomatic relief of pain, use high THC strains with limonene and myrcene. Other terpenes with anti-inflammatory properties are beta-myrcene, beta-caryophyllene, pinene, borneol, eucalyptol, and delta-3-carene.

Best Method of Ingestion for Autoimmune Disorders

Sublingual and swallowed botanic cannabinoids are effective for symptomatic relief of pain from autoimmune disorders, including AIDS and rheumatoid arthritis. Also, vaporization is often quite effective for symptomatic relief.

Cachexia and Appetite Disorders

The ability of cannabis medicines to encourage appetite is part of popular culture. "The munchies" is a phrase known nationwide. Cachexia (also known as wasting syndrome) is characterized by skeletal muscle loss with or without loss of fat tissue. Cachexia is much more than the loss of appetite, as body mass itself is lost and is responsible for the drastic change in the appearance of patients with late-stage cancers and AIDS. 80% of cancer patients suffer from cachexia and 30% die directly from this disease. Currently, there are no FDA-approved drugs are available for treating cancer cachexia, although several synthetic candidates are undergoing clinical trials (Santos et al., 2021).

Cachexia was at one time the least controversial medicinal benefit of cannabis, although not always the best understood. Cannabis is effective in treating cachexia, but some forms of cachexia, especially ones associated with advanced cancer, may not respond. At high doses, cannabis tolerance may develop and the ability of cannabis to treat cachexia will decline.

The endocannabinoid system is the principal regulator of food intake. Cannabinoids stimulate receptors in the hypothalamus and structures in the hindbrain responsible for appetite regulation. "The munchies" are a response to a high dose of cannabis, typically occurring about 90 minutes after ingestion. This may be because larger doses of cannabis exceed the "sweet spot" of appetite stimulation, and the metabolism must wait for the dose to be metabolized. Appetite stimulation requires a very small dose of cannabis. Marinol (a synthetic and FDA-approved form of THC) is used to treat cachexia in small sub-psychoactive doses of 2.5 milligrams before meals.

Botanic Cannabinoids and Terpenes to Treat Cachexia and Appetite Disorders

THC-dominant chemovars are the most used for appetite stimulation and cachexia treatment. Alternative cannabinoids, such as CBD, have provided mixed results. THCV varieties will inhibit appetite and should be avoided. CBDV chemovars are promising but are, as yet, unstudied. High CBG strains may also be of interest. GW Pharmaceuticals has applied for a patent to use CBG for simultaneous agonism of the CB1 and CB2 receptors in cachexia.

Best Method of Ingestion for Cachexia and Appetite Disorders

Both sublingual and swallowed cannabis medicines are effective, and if there are no accompanying lung issues, the vaporization of botanic cannabinoids can be very effective at quickly stimulating the appetite in moderate doses.

Cancer

Botanic cannabinoid medicines have been successfully used to treat nausea and vomiting resulting from chemotherapy. They have also enhanced the effects of prescription opioid pain medication in treating cancer pain. They can stimulate appetite, encourage sleep, reduce anxiety and depression, and lift the spirits of patients undergoing cancer treatment. All of these contribute significantly to the quality of life for those living with cancer. But cancer treatment also attracts dubious and desperate claims of optimism, resulting in claims of cancer cures, for which there is only promising anecdotal evidence. These overstated and false claims have recently extended to cannabis medicines and cancer.

In studies of cancer cells and some animal models, an entourage of botanic cannabinoids has been demonstrated to inhibit tumor growth through a variety of different mechanisms, though this anti-tumor activity has yet to be established in human clinical trials (Śledziński et al., 2018). The effects include suppression of cancer cell signaling mechanisms and inhibition of both blood vessel growths to the tumor and cancer cells, as well as stimulation of programmed cell death of the cancer cell. Epidemiological evidence indicates that an entourage of botanic cannabinoids provides protective effects from developing lung, bladder, brain, and breast cancer, but cancer describes more than 500 diseases in which the cells divide uncontrollably, and it is unlikely cannabis can protect against them all. Cancer is not one disease, and different forms require different treatments. Our understanding of cancer has progressed but examining the molecular and genetic mechanisms underlying these diseases is a complex task, especially to the degree required to effect a cure.

Botanic cannabinoids have been proven to counteract nausea and vomiting resulting from chemotherapy, and they are viewed as a new class of anti-emetic drugs. Pharmaceutical anti-emetics prevent nausea and vomiting but do not increase appetite. Botanic cannabinoids do both. Also, botanic cannabinoid medicines are quite effective at reducing and even eliminating some forms of cancer pain.

Botanic Cannabinoids and Terpenes to Target Cancer

Δ9-Tetrahydrocannabinol (THC) is, of course, critical because of its apoptotic properties (Rieder et al., 2010), but the cannabinoids CBC, CBD, CBDA, CBG, and THCA are also recommended. THCV should probably be avoided due to its appetite-suppressant tendencies. For the terpenes, target strains high in myrcene for its synergistic and anti-mutagenic properties (Surendran et al., 2021). Limonene and linalool have anti-mutagenic properties as well (Fahmy et al., 2022). In addition, myrcene, limonene, and linalool are anti-depressants and are recommended for this reason.

Best Method of Ingestion

Swallowed forms of Full Extract Cannabis Oils (FECO) are generally effective (Backes, 2017). Still, transdermal patches provide an FDA-approved method of intromission with a quicker onset and are more predictable. Vaporization is quite effective, and titration of dose is easily achieved. In Israel, it is not uncommon to see patients vaporizing during chemotherapy sessions.

Dementia

Dementia is a collection of symptoms that gradually reduces the cognitive and functional ability of the brain (Timler et al., 2020). Dementia impinges on memory, intellect, rationality, social skills, and physical functioning (Anand et al., 2018). The indications associated with dementia present themselves in various ways, including depression, frustration, clinginess, forgetfulness, wandering, sexual aggression, hoarding, sleep disturbances, and increased manifestations of challenging behaviors at the end of the day (Kales et al., 2014). Relentless cognitive fluctuations in patients with dementia have been associated with an individual's impaired ability to engage in activities of daily living, including social interactions, and reduced quality of life (Sun et al., 2018). The lingering development and degeneration associated with dementia mandates that the affected individual receive additional support and assistance to remain at home or, ultimately, admission into residential aged-care facilities with 24-hour care.

According to the World Alzheimer Report 2018, about 50 million people worldwide lived with dementia in 2018, and that number is projected to more than triple by 2050 (Peprah & McCormack, 2019). Dementia is the seventh leading cause of death and a major cause of dependency and disability globally, and it is among the top 10 leading causes of years of healthy life lost due to disability in people 60 years old and older (Wimo et al., 2019). The personal consequences of living with dementia are often harrowing for individuals and their families, and the social and economic consequences are challenging for societies and healthcare delivery systems.

The most common type of dementia is Alzheimer's disease, which accounts for about two-thirds of all dementia. Less frequently occurring dementia types include vascular dementia, mixed dementia, Lewy body dementia, frontotemporal dementia, and young-onset dementia (Hillen et al., 2019; Peprah & McCormack, 2019). Neuropsychiatric symptoms are common to all dementia types and manifest as agitation, aggression, wandering, apathy, sleep disorders, depression, anxiety, psychosis, and eating disorders. These behavioral symptoms of dementia pose significant risks of injury to the patients and caregivers, reduce quality of life, and may manifest in extreme distress or depression (Hillen et al., 2019).

Neuropsychiatric symptoms or behavioral and psychological symptoms of dementia impact up to 97% of patients at some point in their illness (Steinberg et al., 2008). These are considered the most pervasive and complex symptoms to manage for relatives and caregivers (Beeri et al., 2002). Behavioral and psychological symptoms of dementia significantly diminish the quality of life of patients, family members, and healthcare functionaries. Managing these symptoms entails correcting somatic and psychiatric factors through the implementation of non-pharmaceutical interventions such as psychological, psychosocial, and behavioral therapies (Cloak & Al Khalili, 2022). Despite these behavioral interventions, synthetic medications are often introduced, generating limited effectiveness and deleterious side effects (Corbett et al., 2012; Sink et al., 2005).

Experimental studies and clinical trials demonstrate the safety of administering botanic cannabinoids to dementia patients. There is also preliminary evidence that botanic cannabinoids alleviate neuropsychiatric symptoms such as agitation, irritability, delusions, and apathy (Bahji et al., 2020; Hillen et al., 2019). A group of Swiss scientists conducted a prospective observational study of 19 patients who received botanic cannabinoids and terpenes recommended by their physician with special authorization from the Switzerland Federal Office of Public Health (Pautex et al., 2022). The two-year study assessed changes in botanic cannabinoid dosages, safety parameters, variations in neuropsychiatric problems, agitation, rigidity, the most abrogating daily activity, and disabling behavior trouble scores. The researchers evaluated the pharmacokinetics of phytocannabinoids, quantified plasma levels, and analyzed enzymatic activity. The study's objective was to examine the efficacy and long-term safety of administering tetrahydrocannabinol and cannabidiol as an additional medicine to a poly-medicated population with severe dementia, evaluate clinical improvements, and collect information on the pharmacokinetics of cannabinoids and potential drug-drug interactions.

While this study requires replication, the results indicate that long-term THC and CBD medicines can be administered safely and demonstrate overall positive clinical improvement in poly-medicated older adults with severe dementia and associated problems. Clinical scores demonstrated marked improvements that were stable over time and a reduction of prescription medications. These findings suggest that botanic cannabinoids and terpenes derived from cannabis demonstrate efficacy in treating older adults with severe dementia and its associated problems.

Botanic Cannabinoids and Terpenes to Target Dementia

THC and CBD are thought to be the most useful phytocannabinoids for targeting dementia, the former for its mood-enhancing properties and the latter for its anti-anxiety effects. Regarding terpenes, myrcene is indicated because it synergizes the other terpenes and changes the permeability of the cell membranes to allow better absorption of the cannabinoids into the brain (Jansen et al., 2019). Linalool should also be considered due to its calming effect (Xu et al., 2021). Limonene may also be appropriate due to its neuroprotectant properties and its tendency to enhance mood (Eddin et al., 2021; Zhang et al., 2019).

Best Method of Ingestion

Oral cannabis preparations are considered best because the effects are long-lasting and easily incorporated into a palatable and appealing form for the Alzheimer's patient. Care should be taken not to leave the oral preparations around where they could be ingested by pets or mistaken by the patient as a snack. Because of its utility related to dose control, FDA-approved transdermal patches provide a highly useful mode of administration.

Diabetes and Obesity

Obesity and diabetes have become health problems of epidemic proportions in the industrialized world (Aronne & Thornton-Jones, 2007; Trillou et al., 2004). More than a million and a half studies have been penned about obesity alone, and the condition is still considered extremely difficult to control in the modern world. Traditional treatments of low-calorie diets and appetite-suppressing drugs frequently fail (Wadden, 1993). Most studies reporting weight loss resulting from merely low-calorie diets report that most subjects regain the weight back either partially or completely within three to five years after treatment ends, and long-term studies present a less favorable outcome, with 49.5% of subjects regaining or surpassing their previous weight (Gosselin & Cote, 2001). These issues with traditional treatment efficacy indicate the necessity for the development of new strategies of both losing weight and maintaining that weight loss.

Nearly half of adult diabetics are considered obese, suggesting that weight loss is an important intervention in the effort to reduce the impact of diabetes on the healthcare system (Nguyen et al., 2007). The total estimated cost of diagnosed diabetes in the U.S. in 2022 was $412.9 billion, including $306.6 billion in direct medical costs and $106.3 billion in indirect costs attributable to diabetes. For cost categories analyzed, care for people diagnosed with diabetes accounts for 1 in 4 healthcare dollars in the U.S., 61% of which are attributable to diabetes. On average, people with diabetes incur annual medical expenditures of $19,736, of which approximately $12,022 is attributable to diabetes. People diagnosed with diabetes, on average, have medical expenditures 2.6 times higher than what would be expected without this disease. Glucose-lowering medications and diabetes supplies account for approximately 17% of the total direct medical costs attributable to diabetes (Parker et al., 2024).

Diabetes is a group of metabolic conditions in which the body does not produce enough insulin or has become resistant to its effects. Insulin is a hormone required to convert sugar, starches, and other food into energy. The two most common forms of diabetes are designated type 1 and type 2. In type 1 diabetes, the pancreas does not produce insulin. Type 2 is a much more common form, normally affecting adults, and is associated with obesity. In type 2 diabetes, the body becomes resistant to the effects of insulin, which enables glucose to accumulate to dangerous levels within the body. High glucose levels damage vascular and other tissues, resulting in heart disease, stroke, blindness, and kidney and nerve damage. Diabetes is the leading cause of preventable blindness among adults. The effectiveness of cannabis in addressing the underlying causes and complications of pre-diabetes and diabetes is, of course, still being researched but is very promising.

Traditional approaches for treating diabetes involve the patient keeping close watch over their blood sugar levels and maintaining them within parameters set by their doctor, and by employing a combination of diet, synthesized medications, and exercise (Preedy, 2020). People with diabetes frequently use complementary and alternative medicine (CAM) techniques of multiple varieties ranging from dietary approaches to herbal and vitamin therapies and massage. The endocannabinoid system has lately received considerable attention as a potential therapeutic target in combating obesity as well as its associated metabolic abnormalities (Richey et al., 2009).

Studies by Sanofi-Aventis Pharmaceuticals demonstrated that a simple synthetic CB1 receptor antagonist (Rimonabant) corrected the deleterious effects of diet-induced obesity by restoring insulin sensitivity and normalizing fat cell size and distribution (Kim et al., 2012). This antagonist also prevented visceral fat accumulation and decreased subcutaneous fat. Other investigations showed similar findings and concluded that blockage of the CB1 receptors with Rimonabant decreased body weight and adiposity, independent of sustained reductions in food intake in humans, canines, and rodents (Jbilo et al., 2005; Kabir et al., 2012; Pi-Sunyer et al., 2006; Richey & Woolcott, 2017; Trillou et al., 2003).

By the year 2006 Sanofi-Aventis had conducted numerous studies which indicated the central cannabinoid (CB1) receptors played a significant role in controlling food consumption and dependence. To develop suitable synthetic medicines against this target, compounds

with potential activity against this receptor were screened for inhibitory activity. Rimonabant emerged from this screening process as a potent CB1 receptor antagonist. Preclinical animal trials subsequently showed that it reduced consumption of fats and sugars, which are significant

contributors to weight gain. These preclinical findings were confirmed in a series of clinical studies involving over 6,000 obese subjects and carried out in both the Americas and Europe. In the United States, the FDA requires two years of safety data before approving anti-obesity medicines, and as part of their patent application process, the pharmaceutical company conducted those trials. The conclusion of the FDA meta-analysis of Rimonabant safety data indicated an increased risk for suicidal ideation in patients, and two suicides were recorded across the two-year Rimonabant clinical trial program. Furthermore, an analysis of data collected from four double-blind, randomized controlled trials demonstrated that 20 mg per day of this synthetic cannabinoid increased the risk of psychiatrically adverse events, specifically depressed mood disorders, and anxiety (Buggy et al., 2011; Christensen et al., 2007; Thomas et al., 2014). These findings resulted in marketing authorization being withdrawn for Rimonabant because the adverse psychological effects could not be addressed (Smaga et al., 2014).

The results of these studies demonstrate that Rimonabant is a synthetic cannabinoid CB1 receptor antagonist that inhibits appetite, regulates blood sugar levels, and reduces the body's resistance to insulin. However, as side effects, this synthetic cannabinoid causes depression and suicidal ideation when ingested. These results beg the question: Would its biologically produced equivalents produce similar results? The endocannabinoid equivalent of Rimonabant is Virodhamine, and the phytocannabinoid equivalent is Δ9-Tetrahydrocannabivarin (THCV). Both biologically produced cannabinoids inhibit appetite, regulate blood sugar levels, and reduce the body's resistance to insulin (Abioye et al., 2020; Porter et al., 2002). However, due to the ban on research of biologically produced cannabinoids in the United States, this has never been studied. Given the proliferation of companies marketing isolate organic phytocannabinoid supplements throughout the nation, studies examining the possible advantageous and deleterious effects of these products on a population are necessary. With 98% of the nation now allowing medicinal botanic cannabinoid use, clinical and policy concerns regarding the mental health effects of biologically produced cannabinoids should be examined now that the Nixon mandate of only studying synthetic cannabinoids has been rescinded. Analyzing this mandate from the paradigm of synthetic cannabinoid/biologic cannabinoid equivalency, it becomes apparent that a survey study of possible depressive properties of THCV and CBD isolates could easily be conducted at suitable research universities. Given the fact that the population is already legally intromitting these supplements, it seems appropriate to study their effects (Dawson, 2018).

Botanic Cannabinoid and Terpenes to Target Diabetes

The endocannabinoid system plays a key role in the development of diabetes and its complications. Diabetic complications linked to the endocannabinoid system include blindness, kidney failure, heart disease, and neuropathic pain. Cannabinoids with reduced or no psychoactivity, including CBD, CBDV, and THCV, are of much interest in maintaining pancreatic function and reducing insulin resistance. Recent research indicates that CBD plays a major role in preventing retinal damage associated with diabetes by acting as an antioxidant and enhancing the retina's own defenses against inflammation (Aebersold et al., 2021; Santiago., 2019).

THCV is also of great interest because of its ability to suppress appetite as well as regulate blood sugar levels and reduce insulin resistance. The THCV botanic cannabinoid is likely to be a major contributor to the prevention and treatment of diabetes. Still, high-THCV cannabis is difficult to find in the United States. High CBD strains are increasingly easy to find. CBDV cannabis is not readily available and may exist in the United States but is as yet unidentified simply because of a lack of lab testing for this particular compound. As far as the terpenes go, it is difficult to go wrong by targeting myrcene, limonene, and pinene (Cox-Georgian et al., 2019).

Best Method of Ingestion for Diabetes

Vaporized cannabis has been shown to ease the neuropathic pain of diabetic subjects. In an article published recently in The Journal of Pain, researchers at the University of California, San Diego, evaluated the effectiveness of vaporized cannabis in comparison with a placebo in 16 subjects with allogenic diabetic peripheral neuropathic pain (Wallace et al., 2015). The authors noted:

This small, short-term, placebo-controlled trial of inhaled cannabis demonstrated a dose-dependent reduction in diabetic peripheral neuropathy pain in patients with treatment-refractory pain. … Overall, our finding of an analgesic effect of cannabis is consistent with other trials of cannabis in diverse neuropathic pain syndromes.

Fibromyalgia

Fibromyalgia is a chronic health condition characterized by widespread, severe musculoskeletal pain that affects an estimated 5 to 7% of the global population. Due to the highly comorbid nature of fibromyalgia, patients with the disorder often respond poorly to the synthetic approach to treatment favored by the paradigm adhered to by the American Healthcare Delivery System (Cameron & Hemingway, 2020). The cause of fibromyalgia remains unknown, but its prevalence is seven times greater for women than for men. Fibromyalgia is a rheumatic disorder similar to arthritis. It is characterized by chronic pain throughout the body, heightened and painful response to pressure, insomnia, morning stiffness, and debilitating fatigue. Multiple factors are involved, including nervous and endocrine system abnormalities, social and genetic factors, and environmental stressors. Traditional pharmaceutical treatments target fibromyalgia symptom relief, but patients' response is typically mixed, and botanic cannabinoids have been demonstrated to be an effective alternative (Khurshid et al., 2021).

Fibromyalgia remains poorly understood. It may be the result of overall central sensitization to pain signaling, a defect in neurotransmitter release, or the obstruction of pathways the body uses to inhibit pain signaling. It may also be the body's response to stress. Some researchers believe that it results from an imbalance of the endocannabinoid system wherein too much anandamide circulates throughout the body, and that is likely a key underlying factor (Smith et al., 2012). Neurologist Ethan Russo (2004) considers fibromyalgia to be an endocannabinoid deficiency disorder. This would provide an explanatory mechanism for why phytocannabinoid supplementation is so effective in its treatment.

Botanic Cannabinoids and Terpenes to Treat Fibromyalgia

Studies on the treatment of fibromyalgia with botanic cannabinoids have thus far only focused on THC and CBD. All conclude that these are safe and effective for the treatment of fibromyalgia pain (Bourke et al., 2022; Cameron & Hemingway, 2020; Mayorga Anaya et al., 2021; Strand et al., 2023). Regarding terpenes, myrcene, limonene, pinene, humulene, alpha, and beta-caryophyllene all exhibit anti-inflammatory properties (Baron, 2018). Savvy medical cannabis patients generally determine which terpenes fit well with their body chemistry and target those cannabis chemovars.

Best Method of Ingestion

Vaporization is arguably the best approach for controlling fibromyalgia pain because the medicinal effects are felt immediately, and patients quickly learn to titrate their own dose. Because oral ingestion of cannabis delivers long-lasting effects, it is popular with fibromyalgia patients. Care must be taken with this approach to avoid over-medicating, as doing so has been demonstrated to increase pain in a University of California study (Wallace et al., 2007).

Gastro-intestinal Disorders

Common folklore asserts that the most common effect of cannabis on the gastrointestinal (GI) tract is the phenomenon known as the "munchies," a term that everyone is familiar with that refers to the food cravings that strike recreational cannabis users (Backes, 2017). The munchies are triggered in the brain, not the stomach or GI tract, and serve as a mechanism to encourage the ingestion of rich, high-fat foods (Kirkham et al., 2002). endocannabinoid system controls or influences all feeding behaviors and gut functions, from feeding to insulin production and fat storage. Consequently, the endocannabinoid system and its receptors are fundamental to how the body acquires and uses energy.

The profusion of cannabinoid receptors located within the gastrointestinal system is the fundamental reason cannabis has been used effectively for gastrointestinal disorders, from vomiting and cramping to pain and inflammatory conditions. Cannabinoids interact with a range of gut receptors beyond just cannabinoid receptors, including the TRPV1 receptors, though their role is not well understood. Botanic cannabinoids have been demonstrated to be effective in treating chemotherapy-induced nausea and vomiting in scores of studies. Because of the widespread occurrence of cannabinoid receptors throughout the GI tract, it is unsurprising that medicinal cannabis produces a range of effective treatments for GI disorders. As the cannabinoid receptors are better understood, there is considerable promise for botanic cannabinoid-based treatments. Still, the endocannabinoid system is complex, and therapies will have to be better understood to avoid the disappointment of rimonabant.

Rimonabant was a diet drug consisting of synthetic virodhamine and designed to block the CB1 receptor to reduce appetite. At one time, this synthetic cannabinoid was marketed as a cure for diabetes. However, it had the side effect of making patients severely depressed, even suicidal. This is yet another example of the problem of using a single synthetic cannabinoid to treat ailments rather than utilizing botanic cannabinoids and employing the entourage effect. Rimonabant was withdrawn as a diet drug in 2008.

Botanic Cannabinoids and Terpenes to Target Gastrointestinal Disorders

Most patients use Indica chemovars to treat gastrointestinal disorders ranging from appetite stimulation to more serious autoimmune disorders such as Crohn's disease. THC has been demonstrated to be effective in reducing spasmodic activity in the intestines, so high THC strains can be quite effective (Di Carlo & Izzo, 2003), and CBD varieties calm gut cramping and inflammation (Hati et al., 2024). Regarding terpenes, beta-caryophyllene is particularly recommended because of its contribution to treating ulcers (Sharma et al., 2016), and it protects the cells lining the GI tract (Francomano et al., 2019).

Best Method of Ingestion

Oral cannabis medicines can be very soothing to the gut if properly prepared. Avoid strong spices and flavorings in cannabis edibles for gastrointestinal conditions. Most patients inhale cannabis for GI disorders. Vaporization of dried flower is arguably a better approach than combustion, and transdermal patches provide an FDA-approved botanic cannabinoid and terpene delivery method.

Glaucoma

In the 1970s, glaucoma became one of the first medical conditions to be cited as justification for a compassionate exception to prevailing federal laws against the medicinal use of botanic cannabinoids, and the United States government has been passing out joints to glaucoma patients for more than four decades. Cannabis as a treatment for glaucoma was first suggested in 1971 in a study that noted that smoking cannabis lowered Intraocular pressure by 25 to 30 percent (Hepler & Frank, 1971). Glaucoma is one of the medical conditions most cited as being effectively treated by medical cannabis, and the actress Whoopi Goldberg (who suffers from glaucoma) always has her vaporizer with her in case she needs to medicate. In the 1970's researchers tried (and failed) to produce eye drops containing THC. They were unable to do so because botanic cannabinoids are not soluble in water. For this reason, cannabis is infrequently recommended by ophthalmologists for glaucoma treatment, but it remains one of the most common treatments for the disease.

Glaucoma is a catchall term that describes a group of diseases that attack the optic nerve and is the leading cause of blindness. High pressure of fluid within the eye results in nerve damage. Primary open-angle glaucoma is the most diagnosed form of glaucoma and has no notable signs or symptoms except the gradual loss of vision. This buildup of pressure in the aqueous humor damages the retinal nerve cells in the eye because the fluid within the eye does not properly move from behind the iris into a small chamber at the front of the eye, where it filters through a spongy tissue before passing into a larger channel and the bloodstream (Backes, 2017). Modulation of the endocannabinoid system has demonstrated potential as a target for the treatment of glaucoma. Intromission of botanic cannabinoids in experimental models has been demonstrated to reduce intraocular pressure and reduce retinal ganglion cell loss, possibly through independent mechanisms. Still, treating glaucoma is challenging for clinicians, not only due to its complexity but also because of all the substances and metabolic pathways involved. Moreover, the target cells in the eye that are involved in the pathogenesis of glaucomatous disease are many and varied. for these reasons, simply reducing intraocular pressure is not sufficient to guarantee a good prognosis for this disease. The use of adjuvants to counter the basic mechanisms underlying the pathogenesis of glaucoma, therefore appears to be a goal to pursue. While cannabinoids were initially exploited in glaucoma solely based on their intraocular pressure reduction properties (Green, 1979), recent evidence suggests that modulation of the endocannabinoid system is also beneficial due to its role in neuroprotection (Cairns et al., 2016).

Botanic Cannabinoids and Terpenes to Target Glaucoma

A randomized, double-masked, placebo-controlled, 4-way crossover study conducted by Tomida (2006) revealed that THC reduces intraocular pressure in glaucoma. CBD does not ease intraocular pressure, but it does offer neuroprotective properties that likely protect retinal ganglia from injury resulting from glaucoma. CBG reduces intraocular pressure, possesses antioxidant, anti-inflammatory, and anti-tumoral activities, and has anti-anxiety, neuroprotective, dermatological, and appetite-stimulating effects. Several findings suggest that research on CBG deserves to be deepened, as it could be used, alone or in association, for novel therapeutic approaches for several disorders, including glaucoma. Several findings suggest that research on CBG deserves to be intensified, as it could be administered alone or in association with other botanic cannabinoids for novel therapeutic approaches for the treatment of glaucoma. CBN promotes neuroprotection and abrogates changes in extracellular matrix protein while normalizing intraocular pressure levels in the eye. This suggests therapeutic potential for CBN in the treatment of glaucoma (Somvanshi et al., 2022). While marketers of natural products promote the primary terpenes to treat glaucoma based on their substantiated anti-inflammatory properties, no credible scientific studies have been conducted in this area.

Best Method of Ingestion

Orally ingested THC is effective for short-term reduction of intraocular pressure due to glaucoma, but tolerance to the benefits of THC typically develops over time. Vaporization is effective for short-term reduction of intraocular pressure (3 to 4 hours), but tolerance to THC's effects builds following sustained usage, reducing its efficacy. Although the application of suitable botanic cannabinoids into the eye, in the form of eye drops, would be an optimal route of administration, such preparations are difficult to develop due to the insolubility of cannabinoid molecules in water. Transdermal patches provide an FDA-approved method of administration, but research in this area is problematic due to political barriers to the pursuit of scientific truth pertaining to the therapeutic potential of botanic cannabinoids in the United States.

HIV/AIDS

The HIV/AIDS crisis of the 1980s and 1990s accelerated the clinical acceptance of botanic cannabinoid medicines when anecdotal evidence indicated efficacy in treating the symptoms of AIDS and the side effects of the synthetic drugs originally used to treat HIV and AIDS. The utilization of botanic cannabinoids to modulate the endocannabinoid system revealed efficacy in treating the wasting syndrome that caused early AIDS patients to lose dangerous amounts of weight while relieving the nausea and appetite suppression side effects of the first FDA-approved synthetic retroviral treatment for AIDS, azidothymidine (AZT).

In the early '90s, Dr. Donald Abrams, the assistant director of the AIDS program for San Francisco General Hospital, observed that his patients benefited from using botanic cannabinoids. At the time, the paradigm to which the American Healthcare Delivery System adhered was that the intromission of botanic cannabinoids suppresses antibody production, immune system functions, and cell proliferation while disrupting the production of pro-inflammatory cytokines, resulting in susceptibility to viral infections such as herpes and Legionnaires' disease (Bredt et al., 2002). Dr. Abrams observed the benefits his AIDS patients who used botanic cannabinoids were experiencing compared to his patients who did not. Thus began his seven-year battle with the U.S. government to attain approval to pursue truth pertaining to the effects of botanic cannabinoids on the immune system of HIV and AIDS-infected patients.

In 1998, Abrams and his team were granted permission to pursue truth pertaining to the effects of botanic cannabinoids on HIV-infected patients. The results of the study demonstrated that botanic cannabinoid ingestion helped AIDS patients eat, diminished nausea, and resulted in weight gain. The findings of this study motivated AIDS activists to align with the early medical cannabis activists to challenge the U.S. government's assertion that botanic cannabinoids possess no medicinal properties.

The modern medical cannabis movement leaped into the national stage from its beginnings as patients' rights issues unfolded during the HIV/AIDS crisis of the 1980s and 1990s. Botanic cannabinoids were found to aid in relieving the wasting syndrome that made AIDS patients lose dangerous amounts of weight. Botanic cannabinoids also relieved the nausea and appetite suppression side effects of the first synthetic retroviral treatment for AIDS. The government tried to ignore or suppress the medicinal use of these nutraceutical molecules, but to no avail, at which point AIDS activists took up the cause. Activists like "Brownie Mary" Rathbun periodically visited the San Francisco General Hospital's AIDS unit to deliver homemade cannabis edibles to patients.

The human immunodeficiency virus (HIV) caused the AIDS epidemic that began in the United States in 1981. According to statistics compiled by Bosh et al. (2021) and amfAR (The American Foundation for AIDS Research), an estimated 1.2 million people in the United States are currently living with HIV. Approximately 13% of these individuals are unaware that they have it.

Botanic Cannabinoids and Terpenes to Target HIV/AIDS

Human immunodeficiency virus (HIV) is the retrovirus that causes AIDS. HIV infection and synthetic antiretroviral therapy independently induce deficiencies in the endocannabinoid anandamide, resulting in HIV-associated neuropathic pain. Phytocannabinoid supplementation of the endocannabinoid deficiency with $\Delta 9$-Tetrahydrocannabinol (THC), the botanic cannabinoid equivalent to anandamide, has been demonstrated in multiple studies to treat HIV-associated neuropathic pain and address nausea associated with synthetic treatments (Aly & Masocha, 2021). Furthermore, cannabidiol (CBD) and Δ9-Tetrahydrocannabinol (THC) are potent antivirals, and for these reasons, targeting these botanic cannabinoids is arguably scientifically justified (Bredt et al., 2002; Sea et al., 2023). Regarding terpenes, all the primary terpenes exhibit potent antiviral properties. This suggests that these nutraceutical molecules should be targeted as well (Sea et al., 2023).

Best Method of Ingestion

While the traditional method of ingestion of botanic cannabinoid molecules is inhalation of combusted flower, no qualified medical professional will recommend setting a substance on fire and breathing the smoke into the lungs. The oral use of botanic cannabinoids by HIV/AIDS patients goes back to the early 1980's. Cannabis was infused into sweets such as brownies, cookies, and candy in hopes of appealing to those with no appetite. Cannabis-infused lollipops can be quite effective in patients having difficulty with solid food, with the added benefit of more rapid absorption of cannabinoids through the tissues of the mouth. Oral cannabis is quite effective for stimulating appetite, increasing the quality of rest and sleep, and long-lasting analgesia in HIV/AIDS patients. Patients report that vaporized cannabis is particularly effective in treating the neuropathic pain associated with HIV/AIDS and its pharmacological treatments (Wilsey et al., 2013).

Insomnia and Sleep Disorders

Insomnia is the inability to fall asleep or to maintain sleep. Botanic cannabinoids and terpenes in cannabis medicines relieve stress and are mildly sedative (Sarris et al., 2020). While anecdotal evidence suggests that an entourage of botanic cannabinoids and terpenes successfully treat multiple sleep disorders, including insomnia, sleep disruption, and sleep apnea, the underlying mechanisms and impact of cannabis on sleep remain somewhat elusive due to the half-century prohibition of research related to botanic cannabinoids by scientists residing in the United States, the limited number of studies conducted by scientists residing in more scientifically progressive countries, and the variability in the design and methods in the few studies that have occurred. Well-designed and controlled studies in this area are essential, particularly as public policy, federal laws, and medical needs related to botanic cannabinoid and terpene use and their public acceptance views rapidly evolve (Choi et al., 2020).

Botanic Cannabinoids and Terpenes to Target Insomnia and Sleep Disorders

Sleep laboratory studies have demonstrated that botanic cannabinoids vary in their ability to sedate or stimulate. THC produces residual sedation, while CBD is wake-promoting (Martin-Santos et al., 2012). Still, CBD is known to reduce anxiety, which can make it easier to fall asleep (Narayan et al., 2022). CBN, which is produced when THC oxidizes over time, has been demonstrated to be synergistically sedative when combined with THC (Russo, 2011). The essential oils (terpenes) are particularly important when it comes to insomnia and sleep disorders. Myrcene and Terpineol are sedatives, and linalool has a calming effect and should arguably be targeted by anyone having sleep issues (Noriega, 2019).

Best Method of Ingestion

For most patients, vaporized cannabis has demonstrated efficacy for insomnia when ingested an hour or so before bedtime. If the patient is waking in the middle of the night, oral botanic cannabinoids and terpenes may be more appropriate due to their longer-lasting effects. Care should be taken not to overmedicate since the stimulating and psychoactive effects may awaken the patient and make sleep difficult, if not impossible (Backes, 2017).

Migraines and Headaches

Botanic cannabinoids have been utilized in a variety of ways and forms for both symptomatic and preventative treatment of headaches and migraines for millennia (Russo, 2001). Botanic cannabinoid medications were eliminated from the medicinal arsenal of the American Healthcare Delivery System when the FDA emerged from the Patent Office in 1906. Upon its inception, it made the institutionalized decision not to evaluate purported medicinal molecules unless a patent accompanied the application for evaluation. Since naturally produced molecules cannot be patented, these nutraceutical molecules were institutionally eliminated from gaining the glorifying label of FDA-approved (Janssen, 1981). Thus, medicinal molecules that the science of medicine relied on for treating disease and mood disorders since the Neolithic Era were eliminated from the medicinal molecular arsenal of the American Healthcare Delivery System. This has resulted in the flawed paradigm of the American Healthcare Delivery System in which the medicines that nature altruistically provides for the treatment of disease and mood disorders became eliminated from America's medicinal arsenal (Crocq, 2020; Janssen, 1981).

Still, patient-reported relief of migraine symptoms has ignited interest in the use of botanic cannabinoid medicines for migraines (Okusanya et al., 2022). In a survey of medical use of botanic cannabinoid medicines in Germany, Austria, and Switzerland, 10.2% of patients with migraine reported self-medicating with botanic cannabinoids (Schnelle et al., 1999). 35% of respondents reported using botanic cannabinoids for headaches/ migraine in a study in the United States and Canada (Leung et al., 2022). However, the five-decade prohibition of research pertaining to the medicinal properties of botanic cannabinoids imposed by American politicians has ensured limited compelling non-anecdotal evidence of their effectiveness in treating migraines.

There is an abundance of observational, anecdotal, and clinical evidence indicating the efficacy of botanic cannabinoids for treating migraine in adults. However, further research is needed to critically assess effective dosing and safety. Cognizant of the surge of interest in botanic cannabinoid and terpene medicines to treat migraines, there is a critical necessity to implement well-designed studies to evaluate the effectiveness and safety of these nutraceuticals for treating adults with migraines.

Botanic Cannabinoids and Terpenes to Target Migraines and Headaches

When searching for a cannabis chemovar to target migraines and headaches, any THC-rich strain is indicated. It is best to avoid CBD dominate strains because CBD may induce or even exacerbate headaches. After migraine onset, heavier indica strains seem effective due to their combination of sedative and analgesic effects (Stith et al., 2020). Regarding terpenes, Beta-caryophyllene is unique because it directly interacts with the endocannabinoid system, specifically the CB2 receptor. In binding to CB2 receptors, beta-caryophyllene exerts potent anti-inflammatory and analgesic effects, reducing both the frequency and severity of headaches and migraine attacks (Baron, 2018). The result of a survey of 50 medical cannabis users conducted by the Substance Abuse and Mental Health Services Administration (2019) revealed that strains of cannabis of high THC/THCA, low CBD/CBDA, and those with predominant terpenes beta-caryophyllene, and myrcene, were most preferred in headache and migraine groups. It also showed that THC/CBD ratios had a 40% improvement and were potentially best used for acute migraine attacks.

Limonene has demonstrated instances of interaction with the body's endocannabinoid system, which plays a crucial role in regulating mood, pain, and inflammation. Furthermore,

Limonene exhibits powerful anti-inflammatory properties, thereby reducing the inflammation often associated with Migraine attacks. Migraines are often triggered or exacerbated by stress and anxiety. Limonene's anxiolytic (anxiety-reducing) properties are believed to alleviate these triggers. Research also suggests that limonene demonstrates the ability to increase serotonin levels in the brain, a neurotransmitter that plays a key role in mood regulation and can influence Migraine onset and severity (Vieira et al., 2018).

The therapeutic properties of linalool are multifaceted, making it a beneficial terpene for migraine relief. It has strong anxiolytic properties, helping to reduce stress and anxiety, which are common m

igraine triggers. Linalool has been demonstrated to promote relaxation and improve sleep quality, which is crucial for recovery during and after a migraine attack.

By working synergistically with other cannabinoids and terpenes, linalool has demonstrated the ability to amplify the overall therapeutic effects of medical cannabis, leading to more comprehensive Migraine relief (Weston-Green et al., 2021).

Best Method of Ingestion

Moderate doses of botanic cannabinoids have demonstrated efficacy in reducing occurrences of migraines. Botanic cannabinoids and terpenes can be administered sublingually or transdermally for faster onset or swallowed for a slower release of THC. Vaporized cannabis is effective for migraine treatment, especially early in the course of the headache (Backes, 2017).

Multiple Sclerosis and Movement Disorders

Multiple sclerosis is a chronic neuroinflammatory disease of the brain and central nervous system. It is characterized pathologically by demyelinating plaques within both grey and white matter, representing a loss of both myelin sheath and supporting oligodendrocytes (Gold & Hartung, 2005). Over time, this demyelination process results in inflammation, progressive neuroaxonal loss, and disability.

Synthetic therapies for multiple sclerosis can be grouped into two categories: disease-modifying and symptomatic. Synthetic disease-modifying therapies aim to decrease the number, severity, and duration of relapses, maintain remission, and slow the progression of the disease. These therapies are usually immunomodulatory and immunosuppressive treatments, such as

interferon beta, Copaxone, fingolimod, natalizumab, and alemtuzumab (Broadley et al., 2014). Symptomatic therapies that purportedly relieve the distressing or disabling symptoms of multiple sclerosis include anticonvulsants for neuropathic pain, anticholinergic drugs for bladder dysfunction and dysphagia, and botulinum toxin injections for spasticity. The use of these symptomatic therapies is limited by their toxicity and iatrogenic effects (Zajicek et al., 2003).

From March 2022 to February 2023, Giossi et al. (2024) conducted a multicenter, cross-sectional survey study. 5620 adult multiple sclerosis patients completed an anonymous online survey with a primary outcome of establishing the estimated prevalence of covert botanic cannabinoid and terpene use of Italian multiple sclerosis patients. The study found that botanic cannabinoid and terpene use is common in multiple sclerosis patients in Italy, with observed prevalence seemingly greater to the general population and often intended for medical use and without disclosure to their treating physician, although with patient-reported clinical benefits.

Botanic cannabinoid and terpene users more frequently reported an improvement in multiple sclerosis-related symptoms than users of synthetic medicines ($p < 0.001$). The difference was also significant when comparing medical cannabis users to current recreational users (p < 0.001). Pain, spasms or tremors, sleep disturbances, anxiety, and sensory symptoms were the most frequently improved symptoms. Current medical cannabis users more frequently reported significant improvements in pain (p = 0.002), spasms or tremors (p = 0.030), sleep disturbances (p = 0.033), and adverse effects of other drugs (p = 0.012). Current medical cannabis users more frequently reported a dose reduction or discontinuation of synthetic medicines for anxiety, sleep, pain, or other conditions. The legal deterrent was deemed relevant for multiple sclerosis patients, as 41.0% of these patients who do not use botanic cannabinoids stated they would use them medicinally if it was legally permissible to do so. This is in line with data from other European countries.

Botanic Cannabinoid and Terpenes to Target Multiple Sclerosis

A variety of botanic cannabinoids yield potent anti-inflammatory and immunosuppressive agents due to their central and peripheral actions on CB1 and CB2 receptors, which mediate different intracellular pathways when activated (Ożarowski et al., 2021). In addition to the effects of the cannabinoids on CB1 and CB2 receptors, their activity on other transmembrane G protein-coupled receptors has a modulating activity on opioid and serotonin receptors (Ingram & Pearson, 2019). THC is deemed to be a psychoactive component and is utilized to treat neuropathic and chronic pain because of its neuroprotective, anti-emetic, and anti-inflammatory properties (Fischedick, 2017). CBD is not an intoxicant, although it does provide medicinal effects on pain and spasticity (VanDolah et al., 2019).

Regarding terpenes, muscle spasticity often experienced by multiple sclerosis patients occurs due to extensive nerve damage in the brain or spinal cord. This affects the voluntary movement controls, resulting in the muscles feeling rigid and heavy. Synergized by the terpene myrcene, limonene provides notable relief, acts as a muscle relaxant, and reduces spasticity. This helps reduce the deterioration of the muscles and aids in reducing neuropathic and inflammatory pain (Eddin et al., 2021). Multiple sclerosis patients are often subject to painful discomfort due to nerve damage that is caused by overactive immune cells (T cells). The excess of T cells results in extended periods of inflammation that can cause significant damage. Humulene has been observed to inhibit the pro-inflammatory cells and modulate this activity, aiding in the prevention of further body tissue degradation (Nuutinen, 2018). The symptoms of multiple sclerosis often result in sleep issues. Muscle spasticity tends to occur more frequently at night as the muscles are not as active during the day. Linalool acts as a sedative agent that potentially helps patients experiencing such symptoms to experience relief. Linalool interacts with the GABA receptors. GABA directs nerve activity in the brain to mitigate feelings of anxiousness. This helps calm patients and assists in managing spasticity (Weston-Green et al., 2021).

Best Method of Ingestion

A placebo-controlled study of 30 multiple sclerosis patients conducted by the University of California Center for Medicinal Cannabis Research examined the effectiveness of inhaled cannabis on multiple sclerosis pain and spasticity (Corey-Bloom et al., 2012). Sixty percent of the participants were also taking synthetic anti-spasticity medications, and 70 percent were undergoing synthetic disease-modifying treatments such as interferon. They continued these treatments during the study. Approximately 80 percent of the participants had used cannabis previously, and 30 percent had used it in the last year. Two-thirds of these participants required mobility aids such as a cane or wheelchair. The patients were divided into two treatment groups and were given either 4 percent THC cannabis or a placebo which they inhaled. The first treatment phase lasted three days, then 11 days off, and the participants were switched over.

Patients were quantified for spasticity, pain, and walking ability and cognitively tested. The results demonstrated a significant reduction in spasticity when botanic cannabinoids and terpenes were administered by inhalation, as compared with the placebo. Pain symptoms were reduced by an average of 50 percent in the botanic cannabinoid and terpene-treated group. This study's most significant limitation was its lack of participants who had never used cannabis.

MRSA

It is not often that a top federal health official for the US Centers for Disease Control (CDC) allows himself to be quoted using words like "nightmare" about an ailment or utter the sentence, "We have a very serious problem, and we need to sound an alarm" (Richards et al., 2013, p. 6). However, this is exactly what happened on March 5th of the year 2013. During a press conference, Dr. Thomas Frieden, Director of the US Centers for Disease Control and Prevention, announced that more than 70 strains of bacteria have become resistant to the "last-resort" family of antibiotics: imipenem, meropenem, doripenem, and ertapenem, while sending a clear signal that the CDC is taking the drug-resistant bacteria problem seriously. As might be expected when discussing bacterial infections, national boundaries are not a deterrent. The pathogenic world operates independently of national borders, and on the heels of that CDC announcement, the United Kingdom's Chief Medical Officer released a 152-page report in which she called antibacterial resistance a "catastrophic threat" that poses a national security risk as serious as terrorism and that unless this resistance is curbed, "We will find ourselves in a health system not dissimilar to the early 19th century, in which organ transplants, joint replacements, and even minor surgeries become life-threatening" (Torjesen, 2013, p. 1).

This section focuses on Methicillin-Resistant Staphylococcus Aureus (MRSA), although the concepts discussed can be applied to nearly all strains of bacterial infections that have

evolved defenses to traditional approaches to treatment and containment. MRSA is quite prevalent in hospitals and nursing homes, where people with weakened immune systems, open wounds, and invasive medical devices like catheters tend to be at greater risk for infections of

nosocomial origin. Hospitalized people and nursing home residents are often immunocompromised and are, therefore, susceptible to bacterial infections of all kinds. It is not uncommon for surgical as well as nonsurgical wounds to become infected with MRSA, and 49 to 65 percent of healthcare-associated *Staphylococcus aureus* infections reported to the National Healthcare Safety Network (NHSN) are caused by methicillin-resistant strains (Wenzel, 2004) resulting in a mortality rate between 28 percent to 38 percent (Gurusamy et al., 2013).

MRSA is being contracted increasingly frequently outside of healthcare institutions and has become quite common in community settings. Particularly common are what are termed community-acquired outbreaks, which are frequently being reported in an increasing number of populations, including children in daycare facilities, prison inmates, patrons of exercise facilities and locker rooms, and military recruits. Groups such as this do not possess the risk factors

traditionally associated with MRSA infection. These include intravenous drug use and recent hospitalization or residence in a care facility but share the common element of being crowded and confined, with poor hygiene practices proliferating. These elements put inhabitants at increased risk of contamination (Dr. David & Daum, 2010).

More than 39 million articles have been penned about techniques for preventing the spread of MRSA and are neatly summarized in an article appropriately titled "Prevent the Spread of MRSA." This article states, "All the tools we need to prevent the spread of MRSA, or any other multidrug-resistant organisms already exist" (Lovato, 2009, p. 26). After making this provocative and hopefully accurate assertion, the author then proceeds to claim that there is no need for special products or technology designed to combat the spread of multidrug-resistant organisms. The author justifies this stance against innovative approaches by citing a 2006

Centers for Disease Control and Prevention publication titled Guidelines for Management of Multidrug-Resistant Organisms in Healthcare Settings. However, the CDC has admitted that these traditional techniques are falling short in combatting the spread of MRSA as well as a plethora of other multidrug-resistant organisms. An analysis of trends of antimicrobial resistance in twenty-three US hospitals between 1996 and 1999 found significant increases in the prevalence of resistant bacterial infections, including MRSA, oxacillin-resistant Staphylococcus

aureus, ciprofloxacin-resistant Pseudomonas aeruginosa, and ciprofloxacin- or ofloxacin-resistant Escherichia coli (Fridkin et al., 2002).

The evidence indicates, and the CDC admits, that traditional approaches are, if not failing, also not succeeding in getting the job done in combatting and containing the spread of these pesky multidrug-resistant pathogens. These bacterial infections are remarkably adaptable and enduring organisms. They appear to reinvent themselves, seemingly outwitting conventional antibiotics. Staphylococcus, especially in its methicillin-resistant form (MRSA), continues to pose to our healthcare facilities, public institutions, and communities an increasing burden of boils, abscesses of all sizes and locations, as well as potentially fatal sepsis, endocarditis, necrotizing fasciitis and pneumonia (Chambers, 2005). Methicillin-resistant Staphylococcus aureus is genetically distinct from other strains of Staphylococcus aureus in that it is responsible for many difficult-to-treat infections in humans because through natural evolutionary mechanisms (horizontal gene transfer, natural selection, multiple drug resistance) it has become immune to beta-lactam, which is the most widely used group of antibiotics. Typical examples of these antibiotics include Penicillin, oxacillin, and methicillin, in which the penams contain a β-lactam ring fused to a five-membered ring, where one of the atoms in the ring is sulfur, and the ring is fully saturated (Dalhoff et al., 2006).

The time it takes to produce new medicines is not the only factor contributing to humanity's difficulty in rising to this evolutionary challenge. Healthcare professionals currently consider Staphylococcus aureus a threat not only in healthcare settings but to the community environment as well. Before the availability of antibiotics, invasive bacterial infections were often fatal, but the introduction of penicillin in the 1940s dramatically improved survival rates.

Although penicillinase-producing strains soon emerged, methicillin and other penicillinase-stable β-lactam agents filled the breach. This success also resulted in a one-molecule attack strategy for combatting bacterial infections. The development of new antibacterial medicines is a slow, arduous process, and the paradigm the pharmaceutical industry has established inhibits success (Fridkin et al., 2005).

MRSA should no longer be considered an exclusively healthcare-associated pathogen as it has become an epidemic within communities. Harbor-UCLA Medical Center conducted a molecular analysis of five available MRSA strains within fifteen months, that revealed community origins. Investigators from the CDC provide compelling evidence from studies conducted in Baltimore, Atlanta, and Minnesota that between eight and twenty percent of MRSA outbreaks originated in community settings (Centers for Disease Control and Prevention, 2017).

Ironically, the emergence of MRSA in community settings has the potential to lead to the pathogen's demise. Communities can employ Complementary Alternative Medicine approaches to health issues that hospitals are institutionally forbidden to allow. Coupled with this, the pharmaceutical company's paradigm restricts hospitals to attacking the bacterium with an individual molecule. Hospitals and healthcare institutions have adopted the same paradigm because it is usually successful. Analogous to Newtonian physics, which is usually successful, a paradigm shift was required when researchers attempted to apply it to subatomic particles. From an evolutionary perspective, societal transformations are now providing the elements necessary for a Kuhnian paradigm shift in pharmaceutical approaches. Traditionally, bacterial infections are attacked pharmacologically through the utilization of a one-molecule approach. Research has repeatedly demonstrated that this attack strategy becomes ineffective over time because bacteria reproduce so rapidly that they evolve genetic mutations capable of defending themselves against the molecule. The dual-pronged protocol being proposed represents a medicinal paradigm shift analogous to the shift from Newtonian physics to quantum mechanics because it entails a biologic rather than synthetic approach to containing bacterial infections, even so-called "superbugs" such as Staphylococcus aureus. This change in paradigm is necessary because pharmacological approaches to containing the spread of drug-resistant bacteria have proved to be largely ineffective.

It might be disconcerting for people to learn that bacterial infections are ubiquitous in society. One of every three people entering a community fitness center carries staph, and two percent carry MRSA. No data showing the total number of people who contract MRSA skin infections in community settings exists (Kavanagh, 2019), but the CDC has acknowledged that traditional methods of preventing transmission must be community-based and that innovative approaches need to be developed within both healthcare organizations and communities. However, traditional methods should not be abandoned simply because the tide is beginning to turn against humanity. This turn is a natural process of evolution, and bacteria have an evolutionary advantage because they reproduce so quickly and, therefore, develop successful genetic modifications at a faster rate than humans. Unfortunately, humanity must rely on its intellect for survival, and at times, the intellect of humans collectively impedes the evolutionary competitions in which humanity is engaged. Our struggle against bacteria has both physiological and psychosocial components and societal struggles are interfering with both. Many of the traditional methods for attacking bacterial infections have value and should not be abandoned, but to fight this war more effectively, adjustments will have to be made to humanity's attack strategies.

Other than medications that are continuously being developed, most of the protocols for containing bacterial infections were developed in the mid-1900s, while some date back to well before the 1840s. Analyzing each of these protocols individually, it is apparent that new attack strategies need to be developed to ensure humanity's survival in this continuous evolutionary process. The protocols humanity employs in the US have been established by the Center for

Disease Control and Prevention, and while they acknowledge the shortcomings, there is an inherent and evidence-based value in each. The CDC recommends contact precautions when the facility considers MRSA to be of special clinical and epidemiologic significance. In most

federally funded hospitals, where MRSA is ubiquitous, mandated protocols include appropriate patient placement, gloving, gowning, patient transport, patient-care equipment, and environmental measures (Appendino et al., 2008). Given that the CDC has now admitted that these protocols are falling short in their goal of preventing the spread of MRSA within hospitals as well as the surrounding community, these protocols should be analyzed individually to determine if modifications or additions might prove more effective.

- Patient placement: The CDC recommends that single-patient rooms be assigned to patients with known or suspected MRSA infections. When single rooms are not available, patients infected with the same strains of MRSA should be housed together.

- Gloving: Gloves should be worn whenever touching the patient's skin or surfaces and articles near the patient (e.g., medical equipment, bed rails).

- Gowning: Proper attire should be worn upon entry into a patient's room or cubicle. Care should be taken to ensure that clothing and skin do not contact potentially contaminated environmental surfaces, which could result in the possible transfer of microorganisms to other patients or environmental surfaces.

- Patient transport restriction: Transport and movement of patients outside of the room only for medically necessary purposes.

- Patient-care equipment and instruments/devices: When possible, use disposable medical patient-care equipment. If common use of equipment for multiple patients is unavoidable, clean and disinfect such equipment before use on another patient.

- Environmental measures: Care should be taken to ensure that rooms of infected patients are cleaned and disinfected at least daily, focusing on frequently touched surfaces such as bed rails, overbed table, bedside commode, lavatory surfaces, doorknobs, and equipment in the immediate vicinity of the patient.

- Pharmacological approaches: Bacterial infections are attacked pharmacologically through the utilization of Penicillin, oxacillin, or methicillin, clindamycin, daptomycin, linezolid (Zyvox), minocycline, tetracycline, trimethoprim-sulfamethoxazole, or vancomycin.

As the CDC has acknowledged, somewhere these protocols are insufficient. If they were not, humanity would not be in the "nightmare scenario" they described. In a war against pathogenic organisms such as this, the only weapon humanity has evolved to fight with is its intellect. Lovato (2009) was likely correct when she so eloquently stated that all the tools we need to prevent the spread of MRSA or any other multidrug-resistant organisms already exist. However, from a psychosocial perspective, the American Healthcare Delivery System has been institutionally mandated not to use the tools. The seven protocol components the CDC provided are fine and brilliant as far as they go. However, it is now necessary for the sake of the future of humanity to do something innovative. Creativity is the essence of science, and creativity and innovation are the tools necessary to defeat drug-resistant bacterial pathogens. Of the seven containment protocols mandated by the CDC, three are open to creative innovation. The most obvious area to begin with is the failure of the pharmaceutical industry in combatting MRSA and other forms of multidrug-resistant bacteria.

Cannabinol, cannabigerol, cannabichromene, cannabidiol, and Δ9-tetrahydrocannabinol all demonstrate potent activity against all Methicillin-resistant Staphylococcus aureus (MRSA) strains of current clinical relevance (Scott et al., 2022). The molecular mechanism of this antibacterial activity is poorly understood because research on phytocannabinoids has been banned for nearly half a century. What is known is that these simple phenols demonstrate significant antimicrobial properties. From a biomolecular perspective, and to stay in compliance with the ban, we are forced to theorize that the resorcinol/benzenediol part of the phytocannabinoids serves as the antibacterial pharmacophore, with the alkyl, terpenoid, and carboxylic appendices modulating its activity. Given the abundance of *Cannabis sativa* strains in America that produce high concentrations of antibacterial phytocannabinoids, this plant represents a viable source of antibacterial agents to address the problem of multidrug resistance in MRSA and other pathogenic bacteria (Neely & Maley, 2000).

Rigorous trials on the use of botanic cannabinoid molecules as systemic antibacterial agents are certainly warranted. However, these trials were forbidden in 1972. Trials cannot be conducted in the United States because all botanic cannabinoid molecules were declared to be Schedule I drugs with no medicinal properties whatsoever. Forty-seven states have rejected this decree, and the remaining states will surely join the rebellion, but most hospitals will likely remain institutionally restricted from utilizing the antibacterial properties inherent in botanic cannabinoid molecules as doing so would jeopardize their federal funding. Major and minor research institutions are subject to the same constraints, making the pursuit of truth in this area problematic.

Many community organizations, such as gyms and fitness centers, where MRSA proliferates, do not receive federal funding and are therefore not subject to the onerous and unreasonable federal restrictions. The proposed protocol described is not designed to treat MRSA, although it will certainly be used to generate a cure when federal law allows. This protocol is merely designed as a two-pronged approach to keep MRSA from spreading within the community. It incorporates and adopts aspects of the protocol components set forth by the CDC while providing innovation in areas deemed suitable. For example, a particularly compelling component of the CDC protocol entails ensuring that the community environment is cleaned and frequently disinfected, focusing on often touched surfaces. The first component of the proposed protocol is comprised of an antibacterial liquid intended to be delivered in an aerosol form akin to Lysol but would likely be supplied in a pump bottle to be more environmentally friendly. Evidence suggests an organic ethyl acetate solution containing antibiotic botanic cannabinoids and terpenes would act as a germicide and disinfect appropriate surfaces. Depending on the community setting, these include tabletops, floors, countertops, lavatory surfaces, doorknobs, and exercise equipment.

The second prong of this protocol can legally be used in federally funded healthcare organizations while still utilizing phytocannabinoids to attack MRSA and other drug-resistant pathogens from a variety of biomolecular fronts. Pathogens are spread by direct contact with the

patient and by touching items that have been contaminated, such as towels, hospital gowns, bed linens, and privacy curtains. One of the most critical aspects of bacterial transfer is the ability of

these infectious bacterial microorganisms to survive on these very common hospital surfaces. Cotton, polyethylene, and polyester allow the pathogens to survive on these fabrics for months. Polyester is usually the material used in hospitals for privacy drapes. These are handled by both patients and staff when they are drawn around the patient's bed. Staphylococcus aureus survives for weeks to months on this fabric, indicating that such screens could serve as reservoirs for these bacteria, making these fabrics essentially vectors for the spread of Enterococcal or Staphylococcal organisms, as a healthcare worker moves from one patient to another, and their scrubs or lab coats contact different patients.

Research has found that hemp fabrics kill bacteria, including pharmaceutical-resistant strains like MRSA. In a study conducted on hemp-rayon fabric composites comprised of 60% hemp and 40% rayon, after the fabric was infected with staph, researchers found that the hemp material killed the staph bacteria at a rate that could only be described as incredible. The material was found to be 98.5% bacteria-free upon measurement after 24 hours. The same material was also infected with Klebsiella pneumoniae (pneumonia). At first measurement, the pneumonia-infected material was 65.1 percent bacteria-free (Belas et al., 2014; Yüksek et al., 2021; Zamora-Mendoza et al., 2022).

The entire dual-pronged approach for preventing the spread of MRSA and other drug-resistant pathogens cannot be implemented in institutions reliant on federal funding because utilizing the antibacterial properties of botanic cannabinoids was federally prohibited in 1972. However, many community organizations exist as privately owned companies and reside in medical cannabis-friendly states. Both prongs of this protocol can be implemented in these. For example, private gyms, golf courses, and exercise centers provide complimentary towels to customers utilizing their locker-room facilities. As has already been discussed, 33 percent of the people entering these facilities carry staph, and two percent carry the MRSA bacteria into these facilities. Providing hemp towels coupled with a phytocannabinoid-based disinfectant regimen provides a method of attacking pathogens within these community settings. Sports organizations like the International Boxing Federation, which routinely lose fighters to staph, could implement both prongs of this protocol, thereby protecting their fighters, who make up a major portion of their financial expenditures.

Staph infections are common in athletes throughout the world. Locker rooms are a breeding ground for dangerous MRSA infections in the NFL. In the same season, both Payton Manning and Tom Brady contracted staph infections. The list of NFL franchises that have battled

MRSA infections is longer than those that have not (Keller et al., 2020). Theoretically, disinfecting locker rooms with antibacterial phytocannabinoids and terpenes in an ethyl alcohol base coupled with hemp uniforms and towels should lessen these outbreaks in states that

are friendly to CBD products. At this time, the only NFL team that could not legally implement this germicide is the Kansas City Chiefs. Every organization could switch to hemp-blend uniforms.

This is an area rife with research opportunities, but with rare exceptions, these opportunities are institutionally mandated to take place outside the United States. The protocol outlined is also economically infeasible because phytocannabinoid isolates are extremely expensive to produce. However, according to an analysis presented at the annual meeting of the International Society for Pharmacoeconomics and Outcomes Research (ISPOR), the annual

nationwide cost to treat hospitalized patients with Methicillin-resistant Staphylococcus aureus (MRSA) infections was estimated in 2005 to be between 3.2 billion and 4.2 billion dollars annually (Deger & Quick, 2009; Siegel et al., 2007; Zinderman et al., 2004).

Botanic Cannabinoid and Terpenes to Target MRSA

All major cannabinoids, including cannabidiol (CBD), THC, cannabigerol (CBG), cannabichromene (CBC), cannabinol (CBN), their derivatives like cannabidiolic acid (CBDA), cannabichromenic acid (CBCA), inhibit MRSA including the epidemic-causing EMRSA 15 and EMRSA 16., while exhibiting greater efficacy than beta-lactam antibiotics, and linezolid, daptomycin, and vancomycin as well (Mahmud et al., 2021). Regarding terpenes, in a study published by the Public Library of Science, alpha-pinene showed antibacterial activity against the antibiotic-resistant Staphylococcus aureus strain (MRSA), killing the pathogen within eight hours and against the bacterium *Campylobacter jejuni* (Kovač et al., 2015). Myrcene inhibits the growth of *Staphylococcus aureus* (Mahmud et al., 2021), as does caryophyllene (Wu et al.,2014).

An overarching concept to be drawn from all this is that the single patentable synthetic molecule approach favored by the pharmaceutical industry fails because pathogens evolve defenses. Bacterial infections like MRSA reproduce very quickly, adapting to the single-molecule attack strategy. The antibacterial botanic cannabinoids and terpenes attack bacterial pathogens from multiple biomolecular molecular fronts, making it very difficult for pathogens to evolve a defense.

Methods of Ingestion to Target MRSA

Methods of ingestion are varied and are dependent on the location and severity of the infection. These include topicals, ingestible oils, edibles, and vaporization.

Muscular Dystrophy

The simplest way of explaining how botanic cannabinoids and terpenes treat the effects of muscular dystrophy is through an illustrative case study of a muscular dystrophy patient. This disabled patient's experiences with the standardized healthcare delivery system provide insight into administratively mandated policies of the American Healthcare Delivery System that violate the principles of biomedical ethics, resulting in established medical protocols that cause disabled intractable pain patients to succumb to an addiction disorder.

There are many disabled patient ethical encounters with the healthcare system, and some, especially those involving end-of-life decisions, are convoluted. This illustrative case study provides insight into the four aspects of biomedical ethics throughout a four-decade timespan. It concerns a disabled patient diagnosed with a form of muscular dystrophy known as Charcot-Marie-Tooth disease (CMT). CMT is a type of peripheral neuropathy that affects the transmission of information between the central nervous system and the rest of the body. Pain symptoms typically begin between the ages of 5 and 25, and the condition is slowly progressive. With this patient, neuropathic pain began around the age of eighteen. This was, coincidently, at the time when institutions of the American Healthcare Delivery System were federally mandated to comply with the principles of biomedical ethics.

Opioids have been regarded for millennia as the most effective drugs for treating neuropathic pain (Rosenblum et al., 2008). The patient was initially prescribed codeine and developed a tolerance to that opiate within the first ten years of the treatment regimen. As was established protocol defined by the healthcare system, his physician switched the opiate to hydrocodone. Hydrocodone worked effectively for several years until the patient developed tolerance and was switched to Oxymorphone. Tolerance to Oxymorphone took place within six months, and Oxycontin was prescribed. By the time tolerance to this drug developed, the patient was 51, and the physicians recommended surgery to place a Fentanyl pump to administer the opiate directly into the spinal cord. Fortunately, the patient had contracted MRSA from prior surgery, and the pain management specialist was unwilling to perform that operation. The patient was then prescribed his final opioid – 500 milligrams of morphine three times daily. That dose caused the patient to suffer a stroke, which resulted in him losing the ability to walk and use his left arm.

At this time, his doctors informed him he had developed tolerance to every oral opiate invented. The protocol now was to place him into hospice as he had less than six months left to live. They assured him he would go out comfortably because they would manage his pain by increasing intravenous morphine, and over the course of the six months, he would stop breathing. The doctors had followed established healthcare system protocol by addicting this disabled patient to opiates for his entire adult life.

At this point, the patient exerted his autonomy by disassociating from the American Healthcare Delivery System and informed his physicians that he was not really interested in dying and would explore alternative medicines. Against his doctor's orders, he terminated his relationship with the healthcare system and began researching nutraceutical alternatives contained in chemovars of *Cannabis sativa.* To the dismay of his physicians, the nutraceuticals treated the neuropathic pain and the withdrawal issues associated with being addicted to opioids for more than four decades.

At this point, the patient had defied medical protocols, and it is when his journey through the healthcare delivery system became problematic. He was not yet officially a medical cannabis patient but was illegally medicating with an illicit substance that the American Healthcare Delivery System viewed as containing the most dangerous molecules humans can ingest. His next step was to become certified to medicate legally with botanic cannabinoids and terpenes in the state where he resided. The pharmaceutical industry had purchased the three healthcare organizations in his state, and each informed their physicians that they were forbidden to recommend medicinal cannabis to their patients. Unaware of this development, the patient went doctor shopping, searching for a physician willing to recommend medicinal cannabis. He met with seven physicians Each visit was covered by his insurance. Seven reasons were given by physicians as to why they were unwilling to recommend the nutraceutical supplement.

- "I'm in favor of medical cannabis, and I am glad this is coming, but I don't have time to learn anything new."
- "I will not certify anyone for medical cannabis until every last bit of evidence comes in. I don't care how long it takes."
- "I've never recommended medical cannabis before, and I am not going to start now."
- "I'm not willing to recommend drugs that are psychoactive and cause brain damage."
- "I'm not willing to recommend medical cannabis for you. What I am willing to do is surgically insert a tube into your spine to pump pain-relieving narcotics directly into your spinal cord."
- "Ingesting cannabis results in brain damage equivalent to being on Bikini Island during a nuclear blast. As a doctor, I will not be a party to that."
- "Even if you proved to me this medicine would save a patient's life, I still wouldn't give it to them."

The patient finally located the only doctor in the State who provided recommendations for medical cannabis. He had to travel six hours to get to the doctor's office and the visit was not covered by his insurance. The physician provided the patient with information on how the State Healthcare System had chosen to manage the medicinal cannabis issue. The State had followed the model of other states that had established medical cannabis programs. Each had designed different methods of claiming they are the most restrictive state regarding medicinal cannabis. The State the patient resided in had chosen to make the application process virtually indecipherable. The State of Minnesota required their physicians to return to school for another year before being permitted to recommend it to patients. The State of Nevada certified organizations with enough money to lobby politicians to grow the *Cannabis sativa* plant to produce medicinal molecules but mandated that they not use the sun.

The state requirement for the patient was to establish a doctor/patient relationship with the recommending physician. This required three $200.00

fifteen-minute visits. The first visit entailed a get-to-know-you session and signing forms so the doctor could obtain the medical records necessary to demonstrate that the patient had a condition qualifying him to medicate with botanic cannabinoid molecules legally. By the second visit, the doctor had received the medical records that proved the patient had a disease that qualified him to be a medical cannabis patient and wrote the recommendation. He explained that this was different than a prescription because no doctor can prescribe a Schedule I drug. On the third visit, the relationship was established, and the physician faxed the recommendation forms to the State. The patient submitted his fingerprints and waited to receive his medical cannabis card in the mail. Once he received it, he continued to medicate illegally because the dispensaries had not yet opened. A few days before the dispensaries opened, the patient received a letter from his insurance company stating they would not cover any part of his hospice bill because he failed to follow the established medical protocol presented to him by his physicians.

This illustrative case study may provide insight into how a disabled individual might develop mistrust in the American Healthcare Delivery System. It certainly speaks to constructs of beneficence, nonmaleficence, and autonomy, although their interpretation is subject to the reader's views regarding natural medicines. It might also be argued that each component of the healthcare system, financing, insurance, payment, and providers, either bent or blatantly violated the principles of biomedical ethics.

Botanic Cannabinoid and Terpenes to Target Muscular Dystrophy

Muscular dystrophy patients experience neuropathic pain, which is efficaciously treated through the utilization of cannabis chemovars with high percentages of THC because this is the botanic cannabinoid that best alleviates neuropathic pain. Even though THC is considered psychoactive by the American Healthcare Delivery System, the Federal Government currently allows doctors to prescribe synthetic THC in the form of Marinol, but Marinol is a poor substitute for whole-plant therapy. CBD is beneficial due to its propensity to treat inflammatory pain. The other primary botanic cannabinoids are also anti-inflammatories, and CBG synergizes their medicinal properties with the added side effect of stimulating brain cell growth. Weight gain can be problematic for people with muscular dystrophy, so THCV prevalent chemovars should be avoided.

Myrcene is an anti-inflammatory with synergistic effects on botanic cannabinoids and terpenes. Limonene is also anti-inflammatory, along with being a mood enhancer. Pinene is purported to promote focus by improving neurobehavioral function and restoring antioxidant enzyme activity to normal levels while decreasing inflammatory factors in the brain. The anti-inflammatory activity of β-caryophyllene has been demonstrated to be comparable to the potency of the synthetic nonsteroidal anti-inflammatory drug phenylbutazone (Basile et al., 1988). Linalool exhibits anti-inflammatory properties as well as providing calming effects and analgesic actions. So, realistically, the best terpene to target is arguably a decision best left to the patient. This is yet another advantage of being a medical cannabis patient. Patients who utilize these nutraceutical compounds are allowed autonomy as to the selection of molecules, the method of ingestion, and the dosing regimen that the pharmaceutical industry is simply unable to provide because doing so typically entails a risk of death.

Best Method of Ingestion for Muscular Dystrophy

Vaporization of dried cannabis flower has been found to be the most effective and rapid mechanism for relaying the active medicinal molecules to the brain, thereby allowing the sufferer to feel immediate relief from pain as well as offering better control over medication levels. At times, muscular dystrophy patients have compromised respiratory systems and utilize edibles, tinctures, or FDA-approved transdermal patch delivery mechanisms. Again, the patient enjoys autonomy as to the choice of cannabis chemovar, as well as the mode of administration.

Parkinson's Disease

The second most common adult-onset neurodegenerative disorder in the United States is Parkinson's disease. The condition is debilitating and presents both motor and non-motor symptoms. Parkinson's disease is caused by a loss of dopaminergic neurons in the substantia nigra. This results in a loss of control of voluntary movements, manifesting as tremors, rigidity, and abnormal slowness of movement, referred to by clinicians as bradykinesia (Tysnes & Storstein, 2017). Pharmaceutical treatment options are scarce and include targeting the dopamine deficiency with levodopa, a synthetic medicine that targets only the motor symptoms of the disorder, failing to inhibit its progression and is associated with side effects, including mental effects such as irritability, anger, hostility, paranoia, insomnia, awakening effects, and uncontrolled, involuntary movements of the face, arms or legs., in addition to anorexia, nausea, and vomiting (Fahn, 1989).

Botanic cannabinoids and terpenes are widely used by the elderly to alleviate the symptoms of neurological and psychiatric disorders such as Parkinson's disease (Suryadevara et al., 2014). About 44% of the population with Parkinson's disease are currently medicating with botanic cannabinoids and terpenes (Kindred et al., 2017). Botanic cannabinoid and terpene users report lower levels of disability compared to non-users (Kindred et al., 2017).

Parkinson's disease is caused by dopamine deficiency. Dopamine is a neurotransmitter that aids in the coordination of nerve and muscle cells involved in movement. Dopamine deficiency results in the hallmark symptoms of Parkinson's, including tremors, rigidity, slowness of movement, and impaired balance and coordination. By the end stages of Parkinson's, patients may lose more than half of the dopamine neurons in the substantia nigra, a region of the basal ganglia in which dopamine neurons should be prevalent. Furthermore, direct laboratory quantification performed on untreated Parkinson's disease patients examining the cerebral spinal fluid demonstrated a doubling of anandamide levels over age-matched controls ($p < 0.001$), irrespective of the disease stage (Pisani et al., 2005). This suggests a compensatory mechanism in the striatum of Parkinson's disease patients as the body's effort to alleviate dopamine depletion. The results of this study were subsequently supported by another (Kreitzer & Malenka, 2007), the first to demonstrate the endocannabinoid system's role in synaptic long-term depression in motor circuits in Parkinson's disease. These motor deficits present in rodents with dopamine lesions were reversed by combining a dopamine receptor agonist with an endocannabinoid reuptake inhibitor. This finding suggests that progressive dopamine loss in Parkinson's disease in striatal circuits may decrease endocannabinoid tone and that the elevations in anandamide in Parkinson's disease patients are attributable to the body's attempt to compensate for this loss. Furthermore, it provides an explanation of how nutraceutical modulation of the endocannabinoid system through the intromission of phytocannabinoids alleviates symptoms of Parkinson's disease (Paes-Colli et al., 2022).

Endocannabinoids anandamide and 2-arachidonoyl glycerol regulate memory, pleasure, concentration, thinking, movement, concentration, sensory and time perception, appetite, and pain (Babayeva et al., 2016).

Botanic Cannabinoids and Terpenes to Target Parkinson's Disease

The botanic cannabinoids responsible for treating the symptoms of Parkinson's disease are Δ9-Tetrahydrocannabinol (THC) and Cannabidiol (Murase et al., 2014; Shen & Thayer, 1999). THC and CBD both exhibit anticonvulsant properties (Babayeva et al., 2014) and alleviate motor disorders by reducing motor impairments and neuron degeneration (Croxford, 2003). Additionally, these botanic cannabinoids have been demonstrated to be effective in preclinical studies involving excitotoxicity, oxidative stress, neuroinflammation, and motor complications associated with Parkinson's disease (Fernández-Ruiz et al., 2013).

Regarding terpenes, limonene, terpinene, pinene, octanol, linalool, beta-caryophyllene, myrcene, and humulene are arguably efficacious due to their antioxidant and anti-inflammatory properties. Pinene may help overcome some of the adverse effects THC is purported to have on memory. Linalool, myrcene, and humulene may be beneficial due to their sedative effects.

Best Method of Ingestion to Target Parkinson's Disease

Vaporized, sublingual, and transdermal methods of ingestion of botanic cannabinoids and terpenes are the primary delivery techniques for most Parkinson's disease patients. These methods are arguably promising.

Post-Traumatic Stress Disorder (PTSD)

PTSD, or posttraumatic stress disorder, is an incapacitating and life-changing condition characterized by a persistent maladaptive reaction after exposure to severe psychological stressors. Experiencing traumatic events, such as violent personal assaults, natural or human-made disasters, serious traffic collisions, or military combat, can lead to the development of PTSD (North et al., 2016). Individuals who suffer from the condition often re-experience their ordeal through nightmares, flashbacks, panic attacks, anxiety, overwhelming emotions, hypervigilance, detachment from loved ones, and sometimes even suicidal behavior (Ehlers et al., 2004). PTSD has proven to be a complex condition to treat with synthetic medicines. No pharmaceutical treatment has been developed that successfully treats PTSD, and there is consensus among clinicians that current pharmaceutical medicines do not work (Detweiler et al., 2016; Dunlop et al., 2014; Hoskins et al., 2015).

Not all cases of post-traumatic stress disorder share similar underlying mechanisms, and the biomolecular and behavioral mechanisms are poorly understood (Hori & Kim, 2019; María-Ríos & Morrow (2020. The disparate mechanisms make the single-molecule approach favored by the pharmaceutical industry unlikely to achieve symptom mitigation in most cases (Sareen, 2014). Compounding this, data compiled by the Board on the Health of Select Populations, & Committee on the Assessment of Ongoing Efforts in the Treatment of Posttraumatic Stress Disorder (2014), psychotherapy approaches, such as cognitive processing therapy, prolonged exposure therapy, and front-line pharmaceutical treatments for military-related PTSD by the Departments of Veterans Affairs and Defense, do not work for up to 66% of patients. Although trauma-focused and non-trauma-focused interventions may improve symptoms, the ignorance or disregard of the psychophysiological mechanisms causing symptom improvement results in a significant percentage of patients continuing to meet the criteria for PTSD after treatment (Watkins et al., 2018).

Posttraumatic Stress Disorder is a disabling psychological condition among military personnel, veterans, and a significant segment of the civilian population. About 3% of the adult population has PTSD at any given time (McManus et al., 2009). Lifetime prevalence is between 1.9% and 8.8% (Bisson et al.,2015), but this rate doubles in populations affected by conflict (Steel et al., 2009) and reaches more than 50% in survivors of rape (Kessler et al., 1995). Post-traumatic stress disorder has also been considered one of the potential long-term consequences of COVID-19, as the lingering feelings of stress and anxiety appear to be ubiquitous (Schou et al., 2021; Bajoulvand et al., 2022). This would indicate a critical need for innovative treatment strategies for PTSD, as pharmacological and trauma-focused or non-trauma-focused psychotherapy approaches have been demonstrated to be largely ineffective ((Detweiler et al., 2016; Dunlop et al., 2014; Hoskins et al., 2015).

It could be reasonably argued that the most significant problem of the psychotherapeutic and pharmacological protocols is that neither adequately address the effects major stressors have on the biomolecular mechanisms that cause mood disorders that manifest themselves as symptoms of PTSD. This missing knowledge or disinterest in the systemic mechanistic biopsychological changes that occur due to significant stressors contributes to the persistence of the problem. The Veterans Affairs Department reports that the military loses more soldiers to suicide than to combat, indicating that alternative treatment options need to be explored (Maguen et al., 2012).

Since the ratification of the Farm Act in 2018, there has been a growing interest in the therapeutic potential of nutraceutical medications that target the immune and endocannabinoid systems to alleviate the symptoms of PTSD. The behavioral effects of endocannabinoid system medications and their ability to modulate neuroinflammation and oxidative stress make targeting the nervous, immune, and endocannabinoid systems potentially relevant. Multiple research studies demonstrate that people with PTSD who use phytochemicals experience more relief from their symptoms than antidepressants such as fluoxetine, paroxetine, sertraline, venlafaxine, and other psychiatric medications (Orsolini et al., 2019). Data from a 2019 study of veterans with PTSD indicates that medicinal cannabis provides better life quality, fewer psychological symptoms, and reduced use of opiates, alcohol, tobacco, and pharmaceutical medicines (McNabb et al., 2020).

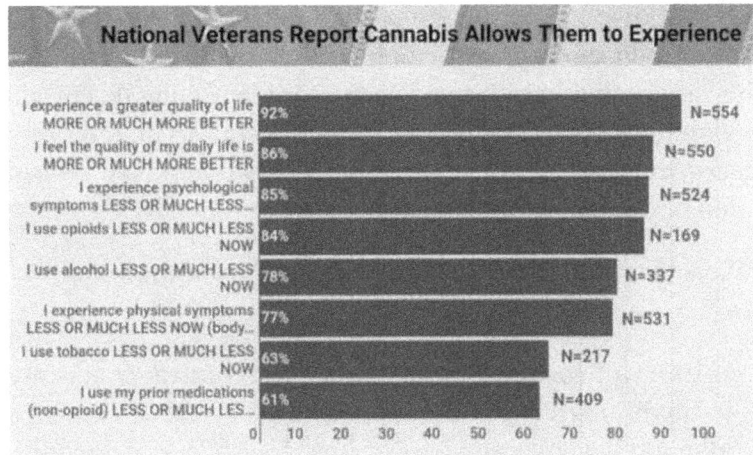

Data from the 2019 Veterans Health and Medical Cannabis study conducted by Dr. Marion McNabb et al.

Still, these studies often fail to address the biomolecular mechanisms that cause these results. This disquisition theorizes that two mechanisms work concomitantly to produce these outcomes. The phytochemicals within cannabis chemovars may be used to modulate critical systems within the body that regulate neurotransmitters in the nervous system, resulting in an efficacious approach to treating the symptoms of PTSD.

The Controversy of Cannabis Chemovars

Theoretical frameworks summarizing theories based on clinical observations and evidence provide guidance in describing the biopsychological mechanisms that cause the nutraceutical approach to surpass pharmaceutical interventions in treating PTSD in terms of efficacy and mitigation of overall symptomology. The pathophysiological foundations of PTSD are complex, and not all cases of the disorder share similar underlying mechanisms (Pitman et al., 2012). The disparate mechanisms make the single-molecule approach favored by the pharmaceutical industry unlikely to achieve symptom mitigation, and in most cases, nutraceutical medicines more effectively treat the symptoms of PTSD sans the side effects ubiquitous through the ingestion of pharmaceuticals (Orsolini et al., 2019). Still, studies that demonstrate that the nutraceutical approach is more efficacious than the pharmacologic typically fail to identify the biopsychological mechanisms that result in these differences. Psychosocial overviews catalyzed investigations into clinical observations deemed useful for identifying this mechanism and devising a nutraceutical protocol expected to match or surpass pharmaceutical interventions in treating PTSD in terms of efficacy and mitigation of overall symptomology.

A Psychosocial and Historical Analysis of the Cannabinoid Sciences

The history of science is remarkably convoluted, especially with regard to the biological molecules produced by the *Cannabis sativa* plant, called phytocannabinoids. On June 18, 1971, President Richard M. Nixon declared war on drugs, explicitly targeting the biological molecules the cannabis plant produces. In 1973, he established the Drug Enforcement Administration (DEA), a paramilitary division of law enforcement tasked with ensuring that no research of biologically produced cannabinoid molecules occurred within the confines of the United States of America (Stone & Robert, 2020). In the late 1980s and 1990s, because the research of these controversial molecules was prohibited in America, the National Institute of Health provided funding to a trio of researchers at Hebrew University in Jerusalem. They utilized newly designed technology to discover the endocannabinoid system. They described this biological system as comprised of endogenous lipid-based retrograde neurotransmitters that bind to cannabinoid receptors and cannabinoid receptor proteins expressed throughout the vertebrate central and peripheral nervous systems (Devane et al., 1992). This group identified the first endogenous cannabinoid neurotransmitter molecule and gave it the name anandamide. In 1993, they discovered a second endocannabinoid molecule, 2-Arachidonoylglycerol (Mechoulam et al., 1995).

On August 11, 2016, Chuck Rosenberg, the acting head of the U.S. Drug Enforcement Administration (DEA), announced it would not change the phytocannabinoids' federal legal status. These controversial molecules would continue to be classified as Schedule I drugs under the Controlled Substances Act, meaning that scientists residing in the United States could not study their potential medical properties. He resigned as head of the DEA in September 2017 after opining that President Donald Trump had little respect for the law (Schmidt, 2017).

The prohibition on the research of cannabinoid molecules lasted 47 years and substantially ended on December 21, 2018, when Donald Trump ratified the Farm Bill. This bill legally reclassified phytocannabinoids extracted from *Cannabis sativa* composed of less than 0.3% THC (hemp) as an agricultural product rather than a controlled substance, thereby legally (not scientifically) differentiating them from the molecules produced by cannabis varieties with higher THC content (Lehrer, 2020). When Donald Trump signed this bill into law, interstate transportation of the known phytocannabinoid molecules became permitted, provided they originated from *Cannabis sativa* classified as hemp. This concession by the U.S. opened the door to the research of biologically produced cannabinoids, which scientists in the United States had been prohibited from investigating for nearly half a century.

Effects of Prohibiting Scientific Inquiry

While the prohibition of research on noncontroversial phytocannabinoids has been rescinded, the paradigm to which cannabinoid scientists have been mandated to adhere remains (Dawson, 2019a). The paradigm was established in 1972 and is currently in what Thomas Kuhn (1962) would have referred to as a state of "crisis" with respect to the potential therapeutic properties of phytocannabinoids. The dominant paradigm requires accepting the proposition that synthetic cannabinoids possess medicinal properties and botanic cannabinoids are dangerous, with none (Drug Enforcement Administration, Department of Justice, 2017). To ensure their freedom and economic security, physicians, clinicians, and cannabinoid scientists capitulated and promoted this view for almost half a century.

Until the passage of the Farm Act, studies that might challenge this paradigm were deemed illegal. This five-decade period defined the science of cannabinoids and dictated the methods of solving puzzles that arose. Ironically, this period was a phase Thomas Kuhn would have referred to as "normal" science. The established paradigm dictated how observational data was perceived, experiments were designed, and the results interpreted. With the methods in place and the assumptions defined, this paradigm flourished. An accumulation of anomalies, such as deaths and other adverse events resulting from synthetic cannabinoids, has resulted in challenges to the dominant paradigm. Kuhn proffered the idea that the scientist's role is to design studies

with the potential to produce results that challenge the dominant paradigm and coined the word "revolution" to describe dramatic changes in scientific worldviews. Revolutionary science is torturous and painful because it shakes all confidence that science has in its present theories and underlying assumptions (Kuhn, 1962). Paradigm shifts occur gradually during a stage when the dominant paradigm is termed to be in "crisis." A new paradigm is now emerging, which professes that phytocannabinoids have medicinal properties without the adverse effects so prevalent through the ingestion of synthetic medicines. With State regulation changes and the Farm Act's implementation, exploring the limitations of the dominant paradigm is now possible. This is the nature of science. When excessive anomalies appear that current theories cannot explain, a period of "crisis" results, and political and economic events fuel the search for new understandings (Dawson, 2019a). This is the stage we are in concerning phytocannabinoid-based medicines.

The recent COVID pandemic has reinforced the precedent established by Harry Anslinger that in the United States, politicians are elected to make medical decisions for the citizens they govern (Dawson &Persad, 2022). Richard Nixon established the precedent that politicians dictate which aspects of truth scientists are permitted to pursue, and scientists have been historically willing to conform and rarely challenge the will of politicians. The five-decade complicity of physicians and clinicians in avoiding searching for knowledge pertaining to the medicinal properties of botanic cannabinoids has resulted in a limited understanding of the effects of intromitting phytocannabinoids on the endocannabinoid system and a lack of human data about the effects of phytocannabinoids on the reparation of traumatic experiences (Berardi et al., 2016). Research on the endocannabinoid system is in its infancy, and there is currently no agreed-upon way by which the endocannabinoid system should be targeted in humans (Ney et al., 2019). Only now are studies beginning to be proposed that might produce results that challenge the accepted assumptions. History has demonstrated that "whether the opposition to attaining and disseminating scientific knowledge is politically or religiously motivated, humanity has the potential to ensure this knowledge is eventually acquired" (Dawson, 2019a, p. 11).

This historical analysis illustrates the convoluted nature of the construct of controversial molecules and the political, cultural, psychosocial, and bureaucratic influencers that bring about this convolution. The Byzantine nature of first declaring war on biologically produced cannabinoid molecules continued with phytocannabinoids' classification as Schedule I drugs in the Controlled Substances Act. The United States Drug Enforcement Administration (DEA) classifies chemicals, drugs, and certain substances used to make drugs into five categories or schedules depending upon the abuse or dependency potential and acceptability as medicine. The potential for abuse is the determining factor in the scheduling of a drug. Schedule I drugs are purported to have a high potential for abuse and create severe psychological and physical dependence. The lower the substance appears on the Schedule, the less potential the drug has to be abused. Legal drugs like alcohol and tobacco do not appear on the Schedule, meaning the DEA considers them outside their purview as the paramilitary division tasked with adjusting the Schedule and enforcing it upon the citizenry (Stone & Robert, 2020).

The half-century prohibition of research in the United States on the medicinal properties of biologically produced cannabinoids has forced American cannabinoid scientists to analyze and develop theories based on studies conducted in other countries, yet ironically funded by the National Institutes of Health. The National Institute of Health is a part of the U.S. Department of Health and Human Services and is considered America's medical research agency. It credits itself with "making important discoveries that improve health and save lives" (National Institutes of Health (2019). Theories are inseparable from clinical observations, and the inability to conduct practical research in their country of origin has resulted in American cannabinoid scientists producing tens of thousands of review articles espousing theoretical explanations for the results of clinical trials funded by the United States but in which they were not involved.

The University of California, Berkley, defines theories as broad, concise, coherent, predictive, broadly applicable natural explanations for an extensive range of phenomena, often integrating and generalizing many hypotheses (Bradford & Hamer, 2022). This chapter examines theories related to devising nutraceutical approaches for treating symptoms associated with post-traumatic stress disorder and a host of other physiological and psychological ailments. The theories interrelate and are based on scientific explanations and interpretations of facts. Furthermore, they are based on clinical observations, have observable consequences, and lead to testable predictions related to biomolecular psychology.

The Construct of Psychoactivity and its Relationship to Controversial Molecules

It appears evident that the DEA is unable to cite studies demonstrating any potential harm of intromitting botanic cannabinoids that justifies them retaining their Schedule I drug classification. Justification for them remaining Schedule I has become based on claims that a particular phytocannabinoid is psychoactive, and therefore pleasure-inducing. A compound's psychoactive nature is considered indicative of its potential for abuse. When research universities were mandated to forbid biological cannabinoid study for fear that conducting such research could jeopardize their federal funding, the only known biologic cannabinoids were phytocannabinoids, the cannabinoids derived from the *Cannabis sativa* plant. These biologically produced molecules were classified as Schedule I drugs, the category reserved for the most dangerous substances humans can ingest (Drug Enforcement Administration, Department of Justice. (2017). This classification for the biologically produced cannabinoids worked well for the next two decades because it fit perfectly into the moral model of addiction, which fueled the drug war (Frank & Nagel, 2017). The DEA flourished during this time through asset forfeitures and massive budgeting allocations provided to ensure the Judicial System punished Americans for possessing cannabinoids produced through biological synthetization.

Things became even more convoluted psychosocially in 1992 when the United States funded the trio of researchers at Hebrew University in Jerusalem. This group identified the first endogenous biological cannabinoid (endocannabinoid) and named it anandamide (Devane et al., 1992). Shortly thereafter, the National Institute of Drug Abuse declared anandamide and Trans-Δ^9-tetrahydrocannabinol (THC) to be the same molecule and justified this position with the image below. Suddenly, it could reasonably be argued that every American possessed an illicit substance merely by being alive (NIDA, 2021).

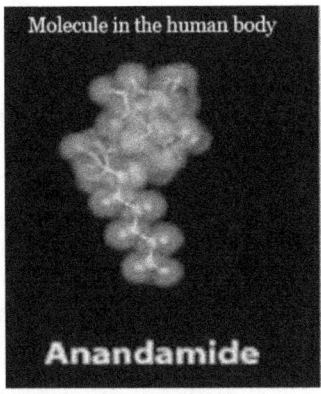 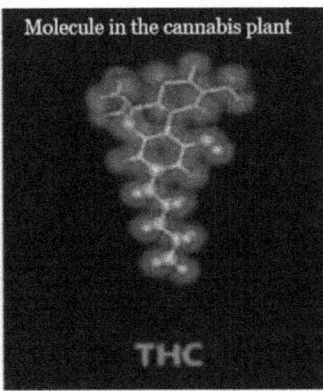

The cannabinoid THC, found in the cannabis plant, is structurally similar to the brain cannabinoid called anandamide. The similarity in structure allows the plant cannabinoid to supplement an anandamide deficiency, thereby altering brain chemistry and treating PTSD. Courtesy of NIDA

Still, the DEA refused to deschedule the botanic cannabinoids and legally categorize them as no-risk drugs like alcohol and tobacco. Instead, politicians admitted the war against the known phytocannabinoids was lost, but hostilities against THC continued, and they misappropriated the term psychoactive. THC would remain the biological cannabinoid with the most destructive potential due to its purported psychoactive nature. This psychoactive property comes from its ability to release dopamine by activating the CB1 receptor in the same way anandamide does, thereby providing a sense of pleasure abhorrent to the moral model of addiction. Cannabidiol was claimed to be nonpsychoactive and, therefore, a noncontroversial botanic cannabinoid. *Cannabis sativa* containing less than 0.3 percent THC was legally classified as hemp, and the phytocannabinoids derived from this plant are legally different from the phytocannabinoids derived from *Cannabis sativa* containing concentrations of THC above the 0.3 percent threshold.

A significant event occurred in biomolecular psychology in the 1970s when the biological mechanism of addiction was identified. It was found that rats would repeatedly and willingly electrically self-stimulate areas in the brain, which were subsequently demonstrated to comprise a set of dopamine neurons (Crow, 1972). This experiment explained the finding of an earlier study, which showed that stimulants enhanced this neurotransmitter's actions (Stein, 1964).

A subsequent series of experiments demonstrated that blocking dopamine receptors with neuroleptic drugs impaired the reinforcing effects of addictive drugs in rats and primates. These studies placed dopamine as the central neurotransmitter in addiction and indicated that it played a critical role in reward, motivation, and incentive behavior (Robinson & Berridge, 1993; Wise & Bozarth, 1987). The conceptual breakthrough coincided with the development of microdialysis sampling techniques pioneered by researchers in Sardinia, Italy. Microdialysis sampling produced conclusive evidence that drugs of abuse release dopamine in the basal forebrain, a preoptic area of the hypothalamus known as the nucleus accumbens. This resulted in a general theory of addiction in which addictive drugs release dopamine, but non-addictive substances do not (Di Chiara & Imperato, 1988). From this point, the field developed rapidly, with replications of the early animal findings of dopamine release by 'addictive' drugs and confirmations in humans using neurochemical imaging. These studies' results led to immense investment in research to alter dopamine neurotransmitter function as a method of treating addiction disorders. Positron imaging tomography (PET scans) and single-photon emission computed tomography (SPECT scans) provided critical breakthroughs in our understanding of the human dopamine system and its role in addiction when it was demonstrated that these technological innovations could be used to measure dopamine release in the human striatum (Laruelle et al., 1996; Volkow et al., 1994). It was later demonstrated that the magnitude of this increase could predict the euphoria or 'high' produced by a drug (Nutt et al., 2015). This proved that in humans, the feeling of pleasure produced by addictive drugs is mediated by striatal dopamine release by the same mechanism as it is in animals, and addiction has come to be viewed as a disorder of the dopamine neurotransmitter system (Di Chiara & Imperato, 1988; Nutt et al., 2015). The dopamine theory of addiction has generated acceptance by biomolecular psychologists because drugs that induce dopamine release repeatedly correlate with feelings of pleasure or euphoria. This sensation of joy or bliss is indicative of psychoactivity. According to the moral model of addiction, psychoactivity should only be induced by legal chemicals like alcohol, tobacco, caffeine, or physician-prescribed pharmaceutical medications (Urban et al., 2010).

The development of the technology capable of analyzing neurotransmitters and applying the results to the dopamine theory of addiction profoundly affected the creation of synthetic drugs designed to target the brain. Pharmaceutical companies used ventral striatal dopamine release assays to estimate the abuse potential of newly synthesized medicines, rejecting compounds if they were determined to induce pleasure, as determined by increased dopamine concentrations (Urban et al., 2010). It might be argued that this was a concession to the moral model of addiction, that anything that results in a pleasurable sensation should be illegal. Still, this view is disconcerting because animal studies conclusively demonstrate that dopamine activity in the ventral striatum is critical to depression resistance (Tye et al., 2013).

The Construct of Psychoactivity

Definitions are critical in science, and psychosocially, the United States, as an institution, has exploited the fact that definitions have little meaning in politics. For example, when the United States patented CBD to treat neurological conditions resulting from concussion or stroke, Alzheimer's, Parkinson's disease, autoimmune diseases, and HIV dementia, it justified the patent application by claiming CBD is non-psychoactive. Psychoactive is an interesting term, and politicians used it to shift the war against *Cannabis sativa* to a confrontation with the Δ9-THC molecule. In 2018, President Donald J. Trump signed what was analogous to an Armistice agreement in the form of the Farm Act, effectively ending the longest war in American history (Trautman et al., 2021). The Farm Act was an institutionalized admission that the United States had lost the war with the plant, but hostilities continued against the phytocannabinoid it claimed to be psychoactive.

As already alluded to, scientists are consummate conformers when politicians determine what they may research, especially when it involves a subject as convolutedly simple as the science of cannabinoids. Recently, a group of scientists published a list of nonpsychoactive phytocannabinoids (Martínez et al., 2020). These consist of cannabidiol (CBD), cannabigerol (CBG), cannabichromene (CBC), and cannabidivarin (CBDV), singling out Δ9-Tetrahydrocannabinol (THC) as psychoactive. The article defines "hemp," "functional foods," "nutraceuticals," and "novel foods" but fails to define the term "psychoactive," the purported subject of the paper. Defining "psychoactive" may have been considered unnecessary by these scientists and their peer reviewers because it is frequently used in the scientific and common vernacular, and everyone intuitively knows what the term means. Still, because this is science, terms of such critical importance require definition. Psychoactivity has a standard scientific definition that is universally accepted. *Substances that, when taken in or administered into one's system, affect mental processes*, e.g., perception, consciousness, cognition, mood, or emotions, are considered to be psychoactive (Reuter & Pardo, 2017). Examples of psychoactive substances include alcohol, caffeine, nicotine, illicit drugs, food, and pharmaceutical medications. In fact, every molecule a human ingests alters brain chemistry, affecting mood, awareness, thoughts, feelings, or behavior, including CBD, CBG, CBC, CBN, CBDV, and H2O (Pitman et al., 2012; Smalheiser, 2019). To claim an individual with PTSD should not ingest a molecule because it may affect mental processes is tantamount to declaring that a person with PTSD should not ingest food. Treatment of many, if not all, physiological and psychological conditions often requires altering brain chemistry, which is the basis of the pharmaceutical and nutraceutical industries.

Theoretical Frameworks toward a Nutraceutical Approach to Treating PTSD

The remainder of this review details theoretical frameworks based on clinical observations deemed useful for devising a nutraceutical protocol expected to match or surpass pharmaceutical interventions in treating PTSD in efficacy and mitigation of overall symptomology. The pathophysiological foundations of PTSD are complex, and not all cases of the disorder share similar underlying mechanisms (Ader & Cohen, 1995). The disparate mechanisms make the single-molecule approach favored by the pharmaceutical industry unlikely to achieve symptom mitigation in most cases. This is an examination of two complementary theories, each reinforcing the other and concluding that a nutraceutical approach incorporating various phytochemicals derived from *Cannabis sativa* chemovars may prove efficacious in mitigating post-traumatic stress disorder symptoms.

The Psychoneuroimmunological Aspects of PTSD

Psychoneuroimmunology is a rapidly growing field of psychology that examines the interactions between the immune and central nervous systems (Ader & Cohen, 1995). The central nervous system and immune systems constantly interact using different biomolecular messengers (Gorman,1997). Different proteins are used by each system. The immune and central nervous systems produce proteins and other molecules that act as messengers between the two systems. The central nervous system uses hormones and neurotransmitters to communicate with the immune system, while the immune system communicates with proteins called cytokines. There are several different types of cytokines, but physiological and psychological stressors activate a kind known as pro-inflammatory cytokines. There are many types of these, and while all are involved in the increase in inflammation, not all these inflammation types manifest themselves within the brain as pathological pain (Liu et al., 2017). Instead, the effect of chronically increased inflammation on the brain causes behavioral symptoms relevant to mood and anxiety disorders, often comorbid (Ong et al., 2018). Some types of chronic inflammation increase systemically over time. This systemic inflammation profoundly affects mood. Greater diversity in day-to-day positive emotions and an elevated sense of well-being are correlated with reducing circulating levels of this type of inflammation (Isung et al., 2020).

Psychoneuroimmunology studies the mechanism of communication between the immune and nervous systems and how this interaction relates to an individual's physiological and psychological well-being. It is fascinating and highly complex and seems to have great potential for treating PTSD, Alzheimer's Disease, and many other types of autoimmune disorders (Cohen et al., 2012).

There is incredible complexity in the emotional states that constitute everyday life, and exposure to psychological trauma and traumatic stressors is related to the onset and perpetuation of specific diseases and various autoimmune disorders (Hori & Kim, 2019). The symptoms of these ailments are considered "side effects" of the inflammatory response that is activated as part of the body's struggle to fight infection. Inflammatory excess is revealed in many patients with PTSD, which may be explained by interactions between the immune and central nervous systems (Schiepers et al., 2005).

Studies with neuroendocrine endpoints have historically dominated research on the biology of PTSD. Still, it is becoming increasingly evident that PTSD psychophysiology is also related to the immune system. Based on the close relationship between the immune and nervous systems and how immune system alterations play a critical role in other neuropsychiatric disorders, it is reasonable to assume that the immune system plays a significant role in the functional symptomology changes that result in the accompanying depression (Pace, 2011).

The pathophysiological foundations of PTSD are complex. As with other psychological disorders, cases of PTSD are not likely to share similar underlying mechanisms (Moraes et al., 2018). Few studies have examined immune system alterations in patients with PTSD, but evidence is accumulating that indicates the immune system plays a pivotal role in its psychopathology and the comorbid psychological symptoms found with the disorder (Dahl et al., 2016; Pace, 2011).

A Nutraceutical Approach to Treating the Psychoneuroimmunological Aspects of PTSD

Studies demonstrate that individuals with PTSD manifest a low-grade inflammatory state characterized by elevated pro and anti-inflammatory cytokine concentrations. Recovery from PTSD is correlated to reducing these levels, irrespective of treatment modality or outcome (Russo & Marcu, 2017; Toft et al., 2020). This correlation indicates that the intromission of anti-inflammatory phytochemicals may help modulate the chronic inflammatory state of PTSD, as quantified by concentrations of inflammatory cytokines (Henshaw et al., 2021). The phytocannabinoids CBD, CBG, CBN, and CBC have been individually identified as anti-inflammatory agents, and the entourage of these molecules enhanced the phytocannabinoid-mediated pro and anti-inflammatory cytokine responses in preclinical in vivo studies. These molecules are noncontroversial and Generally Regarded as Safe (GRAS) by the FDA. An experiment demonstrated that the most commonly studied pro-inflammatory cytokines, tumor necrosis factor-alpha, interleukin (IL)-1b, IL-6, and interferon-gamma, levels are consistently reduced after treatment with CBD & CBG, or CBD & THC, but not with THC alone (Russo, 2008). Since THC is a controversial molecule and its anti-inflammatory potential is only activated in conjunction with another phytocannabinoid, and to stay in compliance with the dominant paradigm, it might justifiably be eliminated from consideration in any potential anti-inflammatory nutraceutical formulation.

Theory of Endocannabinoid Deficiency Disorders

Dr. Ethan Russo proposed the theory of endocannabinoid deficiency disorders. The essence of this theory is that a deficit of certain endocannabinoids is the underlying cause of several physiological and psychological ailments. In a review article published in 2004, Russo cites examples of anandamide being involved in the production of serotonin and pain modulation. He postulated that migraine, fibromyalgia, irritable bowel syndrome, and other clinical conditions exhibit biomolecular and pathophysiological patterns indicating an underlying clinical endocannabinoid deficiency that can be suitably treated through the administration of phytocannabinoid-based medicines (Ibarra-Lecue et al., 2018).

Phytocannabinoid Supplementation for Endocannabinoid Deficiency Disorders

Human and animal studies have consistently demonstrated that the endocannabinoid system is fundamental for physiological homeostasis and proper emotional and cognitive function (Faraone et al., 2005). All vertebrates have been confirmed to possess a measurable endocannabinoid tone reflecting concentrations of anandamide (AEA) and 2-arachidonoylglycerol (2-AG). These are categorized as centrally acting endocannabinoids, and their decreased concentration shows a significant correlation to the development of a variety of physiological and psychological disorders. Furthermore, deficiencies of different endocannabinoid system elements contribute to the pathophysiology of multiple physical and mental disorders, with varying alterations of CB1 and CB2 receptors being demonstrated (Bassir et al., 2019; Skaper & Di Marzo, 2012).

Subsequent research demonstrates that underlying endocannabinoid deficiencies undeniably play a role in migraine, fibromyalgia, irritable bowel syndrome, and an expanding list of other physiological and psychological ailments. Endocannabinoid deficiencies have been demonstrated and theorized to contribute to physiological and psychological disorders, including PTSD (Wilker et al., 2016), postpartum depression (Krumbholz et al., 2013), dementia (Halpern & Walther), diabetes (Dawson, 2018), and ADHD (Asherson & Cooper, 2017). Experimental results confirm the theory, and researchers predict that clinical trials will further establish the efficacy of phytocannabinoid supplementation as a treatment for various endocannabinoid deficiency disorders. As scientific bias fades and legal barriers continue to fall, this will become increasingly apparent (Mayo et al., 2022).

Subversion of the Dominant Paradigm

While pharmacologists and clinicians have frantically been attempting to remain within the bounds of the dominant paradigm by avoiding designing and developing medicines that activate dopamine, thereby producing pleasurable sensations the paradigm deems detestable, animal and human studies demonstrate that CB1 receptor activation within the amygdala is essential for the extinction of fear memories (Elliott et al., 2015; McNabb et al., 2020; Murataeva et al., 2014). Failure to extinguish traumatic memories is a fundamental difficulty in treating post-traumatic stress disorder. This implies that intensifying endocannabinoid signaling has therapeutic utility in treating this condition. Specifically, direct CB1 receptor activation, using Δ9-Tetrahydrocannabinol (THC) or its synthetic analogs or through inhibition of the anandamide-degrading enzyme fatty acid amide hydrolase or the 2-arachidonoyl glycerol (2-AG)–degrading enzyme monoacylglycerol lipase demonstrates efficacy (Bonn-Miller et al., 2022). Δ9-Tetrahydrocannabinol (THC) intromission initiates anandamide production, the endocannabinoid responsible for dopamine activation. Despite the intromission of THC at times resulting in enjoyment, multiple studies indicate the efficacy of treating PTSD patients with cannabis doses containing moderate to high concentrations of THC (Elliott et al., 2015; McNabb et al., 2020; Murataeva et al., 2014).

Applying the Entourage Effect to the Development of Nutraceutical Treatments

The most proliferous cannabinoid researcher was Dr. Raphael Mechoulam, a biochemist that NIH funded at the Hebrew University of Jerusalem. Mechoulam was the first researcher to propose the entourage effect theory when he discovered that 2-linoleoylglycerol, 2-oleoylglycerol, and 2-palmitoylglycerol do not bind to cannabinoid receptors but enhance the binding potentiation of 2-arachidonoylglycerol. This observation inspired Mechoulam to propose that these congenators interact synergistically to improve the binding potential of the endocannabinoid to the primary cannabinoid receptors, theoretically through inhibition of 2-arachidonoylglycerol degradation (Murataeva et al., 2014). Ethan Russo expanded on the theory and extended the concept to phytocannabinoids, theorizing that these compounds enhance, heighten, or mitigate the therapeutic effects of other phytocannabinoids (Russo, 2011).

The evidence supporting the entourage effect is so strong that traditional medicine is moving away from the reductionist approach and towards harnessing the synergy of poly-pharmacological Phyto-combinations that modulate the activity of target networks of underlying disease phenotypes (Mukherjee et al., 2021). Still, subverting dominant scientific paradigms is an arduous process, and as with many scientific constructs, the entourage effect is not universally accepted within the scientific community. Acceptance of the construct is correlated to research funding sources. The pharmaceutical industry promotes a one-patentable-molecule approach to medicine while asserting that the purported synergizing mechanisms described by the entourage are not inherently pharmacologically active and suggests the construct is vaguely defined and fosters misrepresentation and abuse by an unscrupulous and unregulated industry, with little compelling clinical data to support it as a reliable phenomenon predictive of beneficial outcomes. This paradigm's potential subversion depends on future studies' results.

Summary, Conclusion, and Suggestions for Future Studies

There are few pharmaceutical medications with demonstrated efficacy for treating posttraumatic stress disorder, and the pursuit of a synthetic medicine with novel mechanisms of action has stalled (Krediet et al., 2020). A recent meta-analysis comparing the efficacy and acceptability of differing pharmaceuticals for treating adults with posttraumatic stress disorder indicated fluoxetine, paroxetine, sertraline, and venlafaxine should be the recommended drug therapy but expressed uncertainty about the efficacy of these and other non-biodegradable pharmaceuticals for the treatment of the disorder (Huang et al., 2020).

The theories summarized above provide guidance for devising a nutraceutical protocol that would be expected to match or surpass pharmaceutical interventions to treat PTSD in efficacy and mitigation of adverse effects. Given the complexity of the condition coupled with an understanding that not all cases of PTSD share an underlying mechanism, the one-molecule approach to treatment favored by the pharmaceutical industry is unlikely to be successful. A nutraceutical approach allows for incorporating an entourage of biodegradable phytocannabinoid molecules to supplement deficiencies of endocannabinoid concentrations while simultaneously utilizing the antiviral and anti-inflammatory properties of terpenes to treat acute inflammation through the principles of psychoneuroimmunology.

Biochemical occurrences within the brain are at the very root of psychology. What takes place at the biomolecular level is well conceptualized. The role of the immune system and endocannabinoid system in stress-related psychiatric symptoms has been investigated in countless animal and human studies. It is now known that both have a dramatic effect on the manifestation and severity of PTSD symptomology. While more research in this area is necessary, it cannot be disputed that traumatic events deplete endocannabinoid concentrations and that exogenously supplementing resultant deficiencies with phytocannabinoids has demonstrated efficacy (Bassir Nia et al., 2019). Studies still need to be conducted on tolerability, method of administration, optimal dosing, degradation rates, and pharmaceutical interactions of phytochemicals in the brain. Studies on the endocannabinoid system and the psychologically therapeutic properties of Phyto-compounds are the research the FDA is requesting private institutes and universities to conduct. To facilitate these studies, they established the Botanical Safety Consortium. This public-private partnership provides guidelines to assure industry and academic IRBs that psychological research is conducted ethically and safely while providing regulations ensuring that research involving phyto-compounds has no possibility of causing harm to human participants (Lindan & Throckmorton, 2019).

The entourage of phyto-compounds the FDA has certified as safe, coupled with the method of administration the FDA has approved, requires scientific scrutinization. The FDA has reached out to universities and private institutions and requested this research, an occurrence that has not happened since 1971. The demonstrated efficacy of these phyto-compounds in treating PTSD as an endocannabinoid deficiency disorder requires research that meets exacting FDA standards.

As discussed in this summarization of theoretical frameworks, not all cases of PTSD share a similar underlying mechanism. While substantial stressors have been demonstrated to significantly degenerate brain chemistry, they also affect the nervous and immune systems. Both systems interact directly with the brain, which comprises the very root of psychology. Interestingly, phytochemicals that the FDA deems as safe and intromitted in a manner the FDA has determined as acceptable are predicted to demonstrate efficacy in reducing inflammatory processes and levels of pro-inflammatory cytokines, thereby treating the nervous and immune system effects resulting from acute stressors on the chemistry of the brain (Moraes et al., 2018). Further research needs to be conducted on psychoneuroimmunology, the endocannabinoid system, and their relationship to the symptomology of PTSD. Potential studies in these areas are diverse, astonishing, falsifiable, and easy to design.

Botanic Cannabinoids and Terpenes to Target PTSD

High CBD and THC strains are best supplemented with a generous amount of CBG for its synergistic effects. THCV is beneficial because it reduces panic attacks and curbs anxiety attacks in PTSD patients without suppressing emotion. There is an argument that pinene should be avoided since this terpene may reduce the ability of the phytocannabinoids to extinguish aversive memories. The mood-enhancing properties of limonene are likely beneficial, as are the synergistic properties of myrcene.

Best Method of Ingestion

Edibles are popular for reducing dreaming, including the nightmares that sometimes plague some PTSD sufferers. FDA-approved transdermal patches containing a full spectrum of botanic cannabinoids will produce the same benefits by infusing the medicinal molecules directly into the bloodstream, thereby bypassing the liver, where ingested oral medicines are degraded and reduced in effectiveness. Vaporization is beneficial in relieving sudden panic attack symptoms quickly.

Seizure Disorders

Thirty percent of epilepsies do not respond to currently available pharmaceutical treatments (Klein et al., 2024). Worldwide, over 20 million people experience pharmaco-resistant epilepsies, indicating a critical need for novel, effective anti-epilepsy medicines, and botanic cannabinoid-based medicines appear promising. Epileptic seizures have been described as an electrical storm within the brain. These storms typically begin at a single spot where nerve cells begin repeatedly firing together in synchronization. This hyperactivity frequently spreads from that one spot to other areas throughout the brain, causing loss of consciousness and convulsions. It is typical for the person experiencing a seizure to need tens of minutes before becoming coherent again.

Botanic cannabinoids have been used to treat epilepsies since medieval times (Whalley, 2007). Queen Victoria's personal physician deemed cannabis as "the most useful agent with which I am acquainted" in the treatment of "attacks or violent convulsions," that "may recur two or three times in the hour," and claimed that these attacks "may be stopped with a full dose of hemp" (Whalley, 2007, p. 7). Although the intromission of botanic cannabinoids and terpenes has been demonstrated to treat seizure disorders, the mechanism by which they do so remains unclear.

What is known is that the endocannabinoid system is involved in numerous aspects of central nervous system activities and disorders (Di Marzo et al., 2015). The hippocampus is one of the most susceptible regions in the brain vulnerable to seizures lasting more than 5 minutes or 2 or more seizures within a 5-minute period without the person returning to normal between them (Sasaki & Yoshizaki, 2015). One of the most abundant G protein-coupled receptors in the central nervous system is the CB1 receptor, and its concentration is particularly high in the hippocampus (Herkenham et al., 1990). Multiple studies have reported the protective effect of CBD on the hippocampus during seizures (Drysdale et al., 2006; Ryan et al., 2009). Multiple hypotheses have been proposed for the mechanism of action of CBD against epilepsy, but the specific mechanism for its doing so remains unclear (Lu et al., 2023). The mechanism by which botanic cannabinoids inhibit seizure activity likely extends beyond interaction with the cannabinoid receptor CB1 to other receptor systems within the body.

Botanic Cannabinoids and Terpenes to Target Seizure Disorders

CBD molecules have become popular among patients with seizure disorders. A small percentage of seizure sufferers don't respond to CBD but would likely respond to CBDV. However, high CBDV strains are very rare in the United States, and few patients have access to them. As far as the terpenes go, myrcene is probably beneficial because of its synergistic properties as well as linalool due to its calming effect.

References

Abrams, D. I., & Guzman, M. (2015). Cannabis in cancer care. *Clinical Pharmacology & Therapeutics,* (6) https://doi.org/10.1002/cpt.108

Acharya, N., Penukonda, S., Shcheglova, T., Hagymasi, A. T., Basu, S., & Srivastava, P. K. (2017). Endocannabinoid system acts as a regulator of immune homeostasis in the gut. *Proceedings of the National Academy of Sciences of the United States of America, 114*(19), 5005–5010 https://doi.org/10.1073/pnas.1612177114

Abioye, A., Ayodele, O., Marinkovic, A., Patidar, R., Akinwekomi, A., & Sanyaolu, A. (2020). Δ9-tetrahydrocannabivarin (THCV): A commentary on potential therapeutic benefit for the management of obesity and diabetes. *Journal of Cannabis Research, 2*(1), 1-6. https://doi.org/10.1186/s42238-020-0016-7

ACLU. (2021). Over-jailed and untreated. American Civil Liberties Union. Available from https://www.aclu.org/sites/default/files/field_document/20210625-mat-prison_1.pdf

Adams R, Hunt M, Clark JH. (1940) Structure of cannabidiol, a product isolated from the marihuana extract of Minnesota wild hemp. I. *Journal of the American Chemical Society.* 62(1) 196–200. https://doi.org/10.1021/ja01858a058

Ader, R., & Cohen, N. (1995). Psychoneuroimmunology: Interactions between the nervous system and the immune system. *Lancet, 345*(8942), 99. https://doi.org/10.1016/S0140-6736(95)90066-7

Aebersold, A., Duff, M., Sloan, L., & Song, Z. (2021). Cannabidiol signaling in the eye and its potential as an ocular therapeutic agent. *Cellular Physiology and Biochemistry: International Journal of Experimental Cellular Physiology, Biochemistry, and Pharmacology, 55,* 1-14. https://doi.org/10.33594/000000371

Ahmed, M., Boileau, I., Foll, B. L., Carvalho, A. F., & Kloiber, S. (2022). The endocannabinoid system in social anxiety disorder: From pathophysiology to novel therapeutics. *Brazilian Journal of*

Psychiatry / Revista Brasileira De Psiquiatria, 44(1), 81-93. https://doi.org/10.1590/1516-4446-2021-1926

Ahn, K., McKinney, M. K., & Cravatt, B. F. (2008). Enzymatic pathways that regulate endocannabinoid signaling in the nervous system. *Chemical reviews, 108*(5), 1687–1707. https://doi.org/10.1021/cr0782067

Ahn, K. H., Mahmoud, M. M., & Kendall, D. A. (2012). Allosteric modulator ORG27569 induces CB1 cannabinoid receptor high affinity agonist binding state, receptor internalization, and gi protein-independent ERK1/2 kinase activation. *The Journal of Biological Chemistry, 287*(15), 12070-12082. https://doi.org/10.1074/jbc.M111.316463

Aiemsaard, J., Singh, R., Borlace, G. N., Sripanidkulchai, B., Tabboon, P., & Thongkham, E. (2022). Antibacterial activity of cannabis extract (cannabis sativa L. subsp. indica (lam.)) against canine skin infection bacterium staphylococcus pseudintermedius. *ScienceAsia, 48*(3), 348-353. https://doi.org/10.2306/scienceasia1513-1874.2022.053

Alger, B. E. (2013). Getting high on the endocannabinoid system. *Cerebrum: The Dana Forum on Brain Science, 2013*, 14.

Alipour, A., Patel, P. B., Shabbir, Z., & Gabrielson, S. (2019). Review of the many faces of synthetic cannabinoid toxicities. *The Mental Health Clinician, 9*(2), 93–99. https://doi.org/10.9740/mhc.2019.03.093

Allen, D. (2016). Survey shows low acceptance of the science of the ECS (endocannabinoid system) at American medical schools. Outworld. Available from http://www.outwordmagazine.com/inside-outword/glbt-news/1266-survey-shows-low-acceptance-of-the-science-of-the-ecs-endocannabinoid-system

Alves, V. L., Gonçalves, J., L., Aguiar, J., Teixeira, H. M., & Câmara, J. S. (2020). The synthetic cannabinoids phenomenon: From structure to toxicological properties. A review. *Critical Reviews in Toxicology, 50*(5), 359-382. https://doi.org/10.1080/10408444.2020.1762539

Aly, E., & Masocha, W. (2021). Targeting the endocannabinoid system for management of HIV-associated neuropathic pain: A systematic review. *IBRO Neuroscience Reports, 10*(109-118), 109-118. https://doi.org/10.1016/j.ibneur.2021.01.004

Alzu'bi, A., Almahasneh, F., Khasawneh, R., Abu-El-Rub, E., Baker, W. B., & Al-Zoubi, R. M. (2024). The synthetic cannabinoids menace: A review of health risks and toxicity. *European Journal of Medical Research, 29*(1), 49. https://doi.org/10.1186/s40001-023-01443-6

American Psychiatric Association, D. S. M. T. F., & American Psychiatric Association, D. S. (2013). *Diagnostic and statistical manual of mental disorders: DSM-5* (Vol. 5, No. 5). Washington, DC: American psychiatric association.

An, Q., Ren, J., Li, X., Fan, G., Qu, S., Song, Y., Li, Y., & Pan, S. (2021). Recent updates on bioactive properties of linalool. *Food & Function, 12*(21), 10370-10389. https://doi.org/10.1039/d1fo02120f

Anand, A., Khurana, P., Chawla, J., Sharma, N., & Khurana, N. (2018). Emerging treatments for the behavioral and psychological symptoms of dementia. *CNS Spectrums, 23*(6), 361-369. https://doi.org/10.1017/S1092852917000530

Anavi-Goffer, S., Baillie, G., Irving, A. J., Gertsch, J., Greig, I. R., Pertwee, R. G., & Ross, R. A. (2012). Modulation of L-α-lysophosphatidylinositol/GPR55 mitogen-activated protein kinase (MAPK) signaling by cannabinoids. *The Journal of biological chemistry, 287*(1), 91–104. https://doi.org/10.1074/jbc.M111.296020

Anguelov, N. (2018). *From criminalizing to decriminalizing marijuana: The politics of social control*. Lexington Books.

Antshel, K. M., Hargrave, T. M., Simonescu, M., Kaul, P., Hendricks, K., & Faraone, S. V. (2011). Advances in understanding and treating ADHD. *BMC Medicine, 9*(1), 72-83. https://doi.org/10.1186/1741-7015-9-72

Appendino, G., Gibbons, S., Giana, A., Pagani, A., Grassi, G., Stavri, M., Smith, E., & Rahman, M. M. (2008). Antibacterial cannabinoids from cannabis sativa: A structure-activity study. *Journal of Natural Products, 71*(8), 1427-1430. https://doi.org/10.1021/np8002673

Aran, A., Cassuto, H., Lubotzky, A., Wattad, N., & Hazan, E. (2019). Brief report: Cannabidiol-rich cannabis in children with autism spectrum disorder and severe behavioral problems--A retrospective feasibility study. *Journal of Autism and Developmental Disorders, 49*(3), 1284-1288. https://doi.org/10.1007/s10803-018-3808-2

Araújo, J., Cai, J., & Stevens, J. (2019). Prevalence of optimal metabolic health in american adults: National health and nutrition examination survey 2009-2016. *Metabolic Syndrome and Related Disorders, 17*(1), 46-52. https://doi.org/10.1089/met.2018.0105

Armstrong, F., McCurdy, M. T., & Heavner, M. S. (2019). Synthetic cannabinoid-associated multiple organ failure: Case series and literature review. *Pharmacotherapy, 39*(4), 508–513. https://doi.org/10.1002/phar.2241

Arnsten, A., Mazure, C. M., & Sinha, R. (2012). Neural circuits responsible for conscious self-control are highly vulnerable to even mild stress. When they shut down, primal impulses go unchecked and mental paralysis sets in. *Scientific American, 306*(4), 48-53. https://doi.org/10.1038/scientificamerican0412-48

Aronne, L. J., & Thornton-Jones, Z. (2007). New targets for obesity pharmacotherapy. *Clinical*

Pharmacology & Therapeutics, 81(5), 748-752. https://doi.org/10.1038/sj.clpt.6100163

Asherson, P., Buitelaar, J., Faraone, S. V., & Rohde, L. A. (2016). Adult attention-deficit hyperactivity disorder: Key conceptual issues. *The Lancet Psychiatry, 3*(6), 568-578. https://doi.org/10.1016/S2215-0366(16)30032-3

Asherson, P., & Cooper, R. (2017). Treatment of ADHD with cannabinoids. *European Psychiatry, 41*, S55. https://doi.org/10.1016/j.eurpsy.2017.01.031

Aso, E., & Ferrer, I. (2014). Cannabinoids for treatment of alzheimer's disease: Moving toward the clinic. *Frontiers in Pharmacology, 5*, 37. https://doi.org/10.3389/fphar.2014.00037

Atalay, S., Jarocka-Karpowicz, I., & Skrzydlewska, E. (2019). Antioxidative and anti-inflammatory properties of cannabidiol. *Antioxidants (Basel, Switzerland), 9*(1) https://doi.org/10.3390/antiox9010021

Babayeva, M., Assefa, H., Basu, P., Chumki, S., & Loewy, Z. (2016). Marijuana compounds: A nonconventional approach to parkinson's disease therapy. *Parkinson's Disease (20420080), 1*-19. https://doi.org/10.1155/2016/1279042

Babayeva, M., Fuzailov, M., Rozenfeld, P., & Basu, P. (2014). Marijuana compounds: a non-conventional therapeutic approach to epilepsy in children. *J Addict Addictv Disord, 1*(002), 1-9. https://doi.org/10.24966/AAD-7276/100002

Backes, M. (2017). *Cannabis pharmacy: The practical guide to medical marijuana -- revised and updated.* Black Dog & Leventhal.

Bahji, A., Meyyappan, A. C., & Hawken, E. R. (2020). Cannabinoids for the neuropsychiatric symptoms of dementia: A systematic review and meta-analysis. *The Canadian Journal of Psychiatry / La Revue Canadienne De Psychiatrie, 65*(6), 365-376. https://doi.org/10.1177/0706743719892717

Bajoulvand, R., Hashemi, S., Askari, E., Mohammadi, R., Behzadifar, M., & Imani-Nasab, M. (2022). Post-pandemic stress of COVID-19 among high-risk groups: A systematic review and meta-analysis. *Journal of Affective Disorders, 319*, 638-645. https://doi.org/10.1016/j.jad.2022.09.053

Banerjee, E., & Nandagopal, K. (2015). Does serotonin deficit mediate susceptibility to ADHD? *Neurochemistry International, 82*, 52-68. https://doi.org/10.1016/j.neuint.2015.02.001

Banerjee, P. (2015). The harm principle at play: How the animal welfare act fails to protect animals adequately. *University of Maryland Law Journal of Race, Religion, Gender and Class, 15*(2), 361-385.

Barchel, D., Stolar, O., De-Haan, T., Ziv-Baran, T., Saban, N., Fuchs, D. O., Koren, G., & Berkovitch, M. (2019). Oral cannabidiol use in children with autism spectrum disorder to treat related symptoms and co-morbidities. *Frontiers in Pharmacology, 9*, 1521. https://doi.org/10.3389/fphar.2018.01521

Baron E. P. (2018). Medicinal Properties of Cannabinoids, Terpenes, and Flavonoids in Cannabis, and Benefits in Migraine, Headache, and Pain: An Update on Current Evidence and Cannabis Science. *Headache, 58*(7), 1139–1186. https://doi.org/10.1111/head.13345

Barry, C. L., & McGinty, E. E. (2014). Stigma and public support for parity and government spending on mental health: A 2013 national opinion survey. *Psychiatric Services, 65*(10), 1265-1268. https://doi.org/10.1176/appi.ps.201300550

Bassir Nia, A., Bender, R., & Harpaz-Rotem, I. (2019). Endocannabinoid system alterations in posttraumatic stress disorder: A review of developmental and accumulative effects of trauma. *Chronic Stress, 3*https://doi.org/10.1177/2470547019864096

Baswan, S. M., Klosner, A. E., Glynn, K., Rajgopal, A., Malik, K., Yim, S., & Stern, N. (2020). Therapeutic Potential of Cannabidiol (CBD) for Skin Health and Disorders. *Clinical, cosmetic and investigational dermatology, 13*, 927–942. https://doi.org/10.2147/CCID.S286411

Baum, D. (2016, April 1). Legalize it all. *Harper's Magazine, 332*(1991), 22. https://search.ebscohost.com/login.aspx?direct=true&AuthType=sso&db=edsgao&AN=edsgcl.446609248&site=eds-live&scope=site.

Beaumont, T., & Li, Y. (2022). *J. S. Mill's Anti-Imperialist Defence of Empire.* Utilitas, 34(3), 242-261. doi:10.1017/S0953820822000036

Bęben, D., Siwiela, O., Szyjka, A., Graczyk, M., Rzepka, D., Barg, E., & Moreira, H. (2024). Phytocannabinoids CBD, CBG, and their derivatives CBD-HQ and CBG-A induced in vitro cytotoxicity in 2D and 3D colon cancer cell models. *Current Issues in Molecular Biology, 46*(4), 3626-3639. https://doi.org/10.3390/cimb46040227

Beeri, M. S., Werner, P., Davidson, M., & Noy, S. (2002). The cost of behavioral and psychological symptoms of dementia (BPSD) in community dwelling alzheimer's disease patients. *International Journal of Geriatric Psychiatry, 17*(5), 403-408. https://doi.org/10.1002/gps.490

Belas, A. K., Warner, P., & Wang, H. (2014). Antibacterial properties of hemp and other natural fibre plants: A review. *BioResources, 9*(2), 3642-3659. https://doi.org/10.15376/biores.9.2.3642-3659

Beltramo, M., de Fonseca, F. R., Navarro, M., Calignano, A., Gorriti, M. A., Grammatikopoulos, G., Sadile, A. G., Giuffrida, A., & Piomelli, D. (2000). Reversal of dopamine D(2) receptor responses by an anandamide transport inhibitor. *The Journal of Neuroscience: The Official Journal of the Society for Neuroscience,* (9)

Benson, M. J., Anderson, L. L., Low, I. K., Luo, J. L., Kevin, R. C., Zhou, C., McGregor, I. S., & Arnold, J. C. (2022). Evaluation of the possible anticonvulsant effect of Δ^9-tetrahydrocannabinolic acid in murine seizure models. *Cannabis and Cannabinoid Research, 7*(1), 46-57. https://doi.org/10.1089/can.2020.0073

Ben-Shabat, S., Fride, E., Sheskin, T., Tamiri, T., Rhee, M. H., Vogel, Z., Bisogno, T., De Petrocellis, L., Di Marzo, V., & Mechoulam, R. (1998). An entourage effect: Inactive endogenous fatty acid glycerol esters enhance 2-arachidonoyl-glycerol cannabinoid activity. *European Journal of Pharmacology, 353*(1), 23-31. https://doi.org/10.1016/s0014-2999(98)00392-6

Berardi, A., Schelling, G., & Campolongo, P. (2016). The endocannabinoid system and post traumatic stress disorder (PTSD): From preclinical findings to innovative therapeutic approaches in clinical settings. *Pharmacological Research, 111*, 668-678. https://doi.org/10.1016/j.phrs.2016.07.024

Bie, B., Wu, J., Foss, J. F., & Naguib, M. (2018). An overview of the cannabinoid type 2 receptor system and its therapeutic potential. *Current opinion in anaesthesiology, 31*(4), 407–414. https://doi.org/10.1097/ACO.0000000000000616

Bifulco, M., & Pisanti, S. (2015). Medicinal use of cannabis in Europe: the fact that more countries legalize the medicinal use of cannabis should not become an argument for unfettered and uncontrolled use. *EMBO reports, 16*(2), 130–132. https://doi.org/10.15252/embr.201439742

Bisson, J. I., Cosgrove, S., Lewis, C., & Robert, N. P. (2015). Post-traumatic stress disorder. *BMJ (Clinical research ed.), 351*, h6161. https://doi.org/10.1136/bmj.h6161

Blake, K. (2021). Beta-caryophyllene: A review of current research. *Alternative & Complementary Therapies, 27*(5), 222-226. https://doi.org/10.1089/act.2021.29349.kbl

Blasco-Benito, S., Seijo-Vila, M., Caro-Villalobos, M., Tundidor, I., Andradas, C., García-Taboada, E., Wade, J., Smith, S., Guzmán, M., Pérez-Gómez, E., Gordon, M., & Sánchez, C. (2018). Appraising the "entourage effect": Antitumor action of a pure cannabinoid versus a botanical drug preparation in preclinical models of breast cancer. *Biochemical Pharmacology, 157*, 285-293. https://doi.org/10.1016/j.bcp.2018.06.025

Blum, K., Amanda Lih-Chuan Chen, Eric, R. B., David, E. C., Chen, T. J., & al, e. (2008). Attention-deficit-hyperactivity disorder and reward deficiency syndrome. *Neuropsychiatric Disease and Treatment, 2008*, 893-917. Board on the Health of Select Populations, & Committee on the Assessment of Ongoing Efforts in the Treatment of Posttraumatic Stress Disorder. (2014). Treatment for posttraumatic stress disorder in military and veteran populations: Final assessment.

Bonn-Miller, M., Brunstetter, M., Simonian, A., Loflin, M. J., Vandrey, R., Babson, K. A., & Wortzel, H. (2022). The long-term, prospective, therapeutic impact of cannabis on post-traumatic stress disorder. *Cannabis and Cannabinoid Research, 7*(2), 214-223. https://doi.org/10.1089/can.2020.0056

Booth, M. (2005). *Cannabis: A history*. New York: Picador.

Bonini, S. A., Premoli, M., Tambaro, S., Kumar, A., Maccarinelli, G., Memo, M., & Mastinu, A. (2018). Cannabis sativa: A comprehensive ethnopharmacological review of a medicinal plant with a long history. *Journal of ethnopharmacology, 227*, 300–315. https://doi.org/10.1016/j.jep.2018.09.004

Bonnie, R. J. (1974). *The marihuana conviction: A history of marihuana prohibition in the united states.* Charlottesville, Virginia: University of Virginia Press

Bonnie, R. J., & Whitebread, C. H. (1974). *The marihuana conviction: A history of marihuana prohibition in the United States* (Vol. 2). Charlottesville: University Press of Virginia.

Borch, A., & Rantala, V. (2015). Addiction: A highly successful, essentially contested concept. *International Journal of Alcohol and Drug Research, 4*(1), 1-4. https://doi.org/10.7895/ijadr.v4i1.200

Borgonetti, V., Benatti, C., Governa, P., Isoldi, G., Pellati, F., Alboni, S., Tascedda, F., Montopoli, M., Galeotti, N., Manetti, F., Miraldi, E., Biagi, M., & Rigillo, G. (2022). Non-psychotropic cannabis sativa L. phytocomplex modulates microglial inflammatory response through CB2 receptors-, endocannabinoids-, and NF-κB-mediated signaling. *Phytotherapy Research : PTR, 36*(5), 2246-2263. https://doi.org/10.1002/ptr.7458

Borrelli, F., Fasolino, I., Romano, B., Capasso, R., Maiello, F., Coppola, D., Orlando, P., Battista, G., Pagano, E., Di Marzo, V., & Izzo, A. A. (2013). Beneficial effect of the non-psychotropic plant cannabinoid cannabigerol on experimental inflammatory bowel disease. *Biochemical Pharmacology, 85*(9), 1306-1316. https://doi.org/10.1016/j.bcp.2013.01.017

Bosh, K. A., Hall, H. I., Eastham, L., Daskalakis, D. C., & Mermin, J. H. (2021). Estimated annual number of HIV infections — united states, 1981-2019. *MMWR: Morbidity & Mortality Weekly Report, 70*(22), 801-806. https://doi.org/10.15585/mmwr.mm7022a1

Bourke, S. L., Schlag, A. K., O'Sullivan, S. E., Nutt, D. J., & Finn, D. P. (2022). Cannabinoids and the endocannabinoid system in fibromyalgia: A review of preclinical and clinical research. *Pharmacology & Therapeutics, 240*, 108216. https://doi.org/10.1016/j.pharmthera.2022.108216

Bradford, A., Hamer, A. (2022). What is a scientific theory? [Internet]. livescience.com. *Live Science*. Available from: https://www.livescience.com/21491-what-is-a-scientific-theory-definition-of-theory.html

Bradshaw, R. H. W., Coxon, P., Greig, J. R. A., & Hall, A. R. (1981). New fossil evidence for the past cultivation and processing of hemp (cannabis sativa L.) in Eastern England. *New Phytologist, 89*(3), 503-510. https://doi.org/10.1111/j.1469-8137.1981.tb02331.x

Brand, E. J., & Zhao, Z. (2017). Cannabis in chinese medicine: Are some traditional indications referenced in ancient literature related to cannabinoids? *Frontiers in Pharmacology*, https://doi.org/10.3389/fphar.2017.00108

Bredt, B. M., Higuera-Alhino, D., Shade, S. B., Hebert, S. J., McCune, J. M., & Abrams, D. I. (2002). Short-term effects of cannabinoids on immune phenotype and function in HIV-1-infected patients. *The Journal of Clinical Pharmacology, 42*(11), 82S-89S. https://doi.org/10.1177/0091270002238798

Broadley, S. A., Barnett, M. H., Boggild, M., Brew, B. J., Butzkueven, H., Heard, R., Hodgkinson, S., Kermode, A. G., Lechner-Scott, J., Macdonell, R. A. L., Marriott, M., Mason, D. F., Parratt, J., Reddel, S. W., Shaw, C. P., Slee, M., Spies, J., Taylor, B. V., Carroll, W. M., . . . Willoughby, E. (2014). Therapeutic approaches to disease modifying therapy for multiple sclerosis in adults: An australian and new zealand perspective part 1 historical and established therapies. *Journal of Clinical Neuroscience, 21*(11), 1835-1846. https://doi.org/10.1016/j.jocn.2014.01.016

Brown, A. S., & Gershon, S. (1993). Dopamine and depression. *Journal of neural transmission. General section, 91*(2-3), 75–109. https://doi.org/10.1007/BF01245227

Buchman, D. Z., Skinner, W., & Illes, J. (2010). Negotiating the relationship between addiction, ethics, and brain science. *AJOB Neuroscience, 1*(1), 36-45. https://doi.org/10.1080/21507740903508609

Buggy, Y., Cornelius, V., Wilton, L., & Shakir, S. A. W. (2011). Risk of depressive episodes with rimonabant: A before and after modified prescription event monitoring study conducted in england. *Drug Safety, 34*(6), 501-509. https://doi.org/10.2165/11588510-000000000-00000

Busquets-Garcia, A., Gomis-Gonzalez, M., Guegan, T., Agustin-Pavon, C., Pastor, A., Mato, S., Perez-Samartin, A., Matute, C., de la Torre, R., Dierssen, M., Maldonado, R., & Ozaita, A. (2013). Targeting the endocannabinoid system in the treatment of fragile X syndrome. *Nature Medicine, 19*(5), 603. https://doi.org/10.1038/nm.3127

Burstein, S., Levin, E., & Varanelli, C. (1973). Prostaglandins and cannabis: II inhibition of biosynthesis by the naturally occurring cannabinoids. *Biochemical Pharmacology, 22*(22), 2905-2910. https://doi.org/10.1016/0006-2952(73)90158-5

Butler, D., & Callaway, E. (2016). Scientists in the dark after french clinical trial proves fatal. *Nature, 529*(7586), 263-264. https://doi.org/10.1038/nature.2016.19189

Cabrera, C. L. R., Keir-Rudman, S., Horniman, N., Clarkson, N., & Page, C. (2021). The anti-inflammatory effects of cannabidiol and cannabigerol alone, and in combination. *Pulmonary Pharmacology & Therapeutics, 69*, 102047. https://doi.org/10.1016/j.pupt.2021.102047

Cairns, E. A., Baldridge, W. H., & Kelly, M. E. M. (2016). The endocannabinoid system as a therapeutic target in glaucoma. *Neural Plasticity*, 1-10. https://doi.org/10.1155/2016/9364091

Calapai, F., Cardia, L., Esposito, E., Ammendolia, I., Mondello, C., Lo Giudice, R., Gangemi, S., Calapai, G., & Mannucci, C. (2022). Pharmacological aspects and biological effects of cannabigerol and its synthetic derivatives. *Evidence-Based Complementary & Alternative Medicine (eCAM), 1*-14. https://doi.org/10.1155/2022/3336516

Cameron, E. C., & Hemingway, S. L. (2020). Cannabinoids for fibromyalgia pain: a critical review of recent studies (2015–2019). *Journal of cannabis research, 2*, 1-11. https://doi.org/10.1186/s42238-020-00024-2

Canitano, R., & Scandurra, V. (2008). Risperidone in the treatment of behavioral disorders associated with autism in children and adolescents. *Neuropsychiatric Disease and Treatment, 4*(4), 723-730. https://doi.org/10.2147/NDT.S1450

Cao, C., Li, Y., Liu, H., Bai, G., Mayl, J., Lin, X., Sutherland, K., Nabar, N., & Cai, J. (2014). The potential therapeutic effects of THC on alzheimer's disease. *Journal of Alzheimer's Disease, 42*(3), 973-984. https://doi.org/10.3233/JAD-140093

Capuano, A., Scavone, C., Rafaniello, C., Arcieri, R., Rossi, F., & Panei, P. (2014). Atomoxetine in the treatment of attention deficit hyperactivity disorder and suicidal ideation. *Expert opinion on drug safety, 13*(sup1), 69-78. https://doi.org/10.1517/14740338.2014.941804

Carson, A. & Anderson, E. (2016). Prisoners in 2015. US Department of Justice. *Bureau of Justice Statistics*. https://www.bjs.gov/content/pub/pdf/p15.pdf

Castaneto, M. S., Gorelick, D. A., Desrosiers, N. A., Hartman, R. L., Pirard, S., & Huestis, M. A. (2014). Synthetic cannabinoids: Epidemiology, pharmacodynamics, and clinical implications. *Drug and Alcohol Dependence, 144*, 12–41. https://doi.org/10.1016/j.drugalcdep.2014.08.005

Castillo, P. E., Younts, T. J., Chávez, A., E., & Hashimotodani, Y. (2012). Endocannabinoid signaling and synaptic function. *Neuron, 76*(1), 70-81. https://doi.org/10.1016/j.neuron.2012.09.020

Centers for Disease Control and Prevention. (2017) General Information About MRSA in the community. Methicillin-resistant Staphylococcus aureus (MRSA) Basics | MRSA | CDC

Centonze, D., Bari, M., Di Michele, B., Rossi, S., Gasperi, V., Pasini, A., Battista, N., Bernardi, G., Curatolo, P., & Maccarrone, M. (2009).

Altered anandamide degradation in attention-deficit/hyperactivity disorder. *Neurology, 72*(17), 1526-1527. https://doi.org/10.1212/WNL.0b013e3181a2e8f6

Chadi, N., Minato, C., & Stanwick, R. (2020). Cannabis vaping: Understanding the health risks of a rapidly emerging trend. *Paediatrics & Child Health, 25*, 16. https://doi.org/10.1093/pch/pxaa016

Chambers, H. F. (2005). Community-associated MRSA - resistance and virulence converge. *The New England Journal of Medicine, 352*(14), 1485-1487. Chandy, M., Nishiga, M., Wei, T., Hamburg, N. M., Nadeau, K., & Wu, J. C. (2024). Adverse impact of cannabis on human health. *Annual Review of Medicine, 75*, 353-367. https://doi.org/10.1146/annurev-med-052422-020627

Chen, J., Marmur, R., Pulles, A., Paredes, W., & Gardner, E. L. (1993). Ventral tegmental microinjection of Δ9-tetrahydrocannabinol enhances ventral tegmental somatodendritic dopamine levels but not forebrain dopamine levels: evidence for local neural action by marijuana's psychoactive ingredient. *Brain research, 621*(1), 65-70. https://doi.org/10.1016/0006-8993(93)90298-2

Chen, J., Paredes, W., Lowinson, J. H., & Gardner, E. L. (1990). Delta 9-tetrahydrocannabinol enhances presynaptic dopamine efflux in medial prefrontal cortex. *European Journal of Pharmacology, 190*(1-2), 259-262. https://doi.org/10.1016/0014-2999(90)94136-1

Choi, S., Huang, B. C., & Gamaldo, C. E. (2020). Therapeutic uses of cannabis on sleep disorders and related conditions. *Journal of Clinical Neurophysiology, 37*(1), 39-49. https://doi.org/10.1097/WNP.0000000000000617

Chopra, I. C., & Chopra, R. N. (1957). The use of cannabis drugs in india. *Bulletin on Narcotics, 9*, 4–29. https://www.unodc.org/unodc/en/data-and-analysis/bulletin/bulletin_1957-01-01_1_page003.html

Civiletto, C. W., & Hutchison, J. (2021). Electronic vaping delivery of cannabis and nicotine. *StatPearls,* https://www.ncbi.nlm.nih.gov/books/NBK545160/

Clarke, H., Roychoudhury, P., Narouze, S.N. (2021). Other Phytocannabinoids. In: Narouze, S.N. (eds) Cannabinoids and Pain. Springer, Cham. https://doi.org/10.1007/978-3-030-69186-8_12

Clark, V. S. (1929). *History of manufactures in the united states. bd. 1. 1607-1860,* McGraw Hill, NY

Cloak, N., & Al Khalili, Y. (2022). Behavioral and psychological symptoms in dementia. *StatPearls,*

Cohen, S., Janicki-Deverts, D., Doyle, W. J., Miller, G. E., Frank, E.,Rabin, B. S., & Turner, R. B. (2012). Chronic stress, glucocorticoid receptor resistance, inflammation, and disease risk. *Proceedings of the National Academy of Sciences of the United States of America, 109*(16), 5995-5999.

Collin, C., Ehler, E., Waberzinek, G., Alsindi, Z., Davies, P., Powell, K., Notcutt, W., O'Leary, C., Ratcliffe, S., Nováková, I., Zapletalova, O., Piková, J., & Ambler, Z. (2010). A double-blind, randomized, placebo-controlled, parallel-group study of sativex, in subjects with symptoms of spasticity due to multiple sclerosis. *Neurological Research, 32*(5), 451-459. https://doi.org/10.1179/016164109X12590518685660

Cooper, R. E., Williams, E., Seegobin, S., Tye, C., Kuntsi, J., & Asherson, P. (2017). Cannabinoids in attention-deficit/hyperactivity disorder: A randomised-controlled trial. *European Neuropsychopharmacology, 27*(8), 795-808. https://doi.org/10.1016/j.euroneuro.2017.05.005

Corbett, A., Smith, J., Creese, B., & Ballard, C. (2012). Treatment of behavioral and psychological symptoms of alzheimer's disease. *Current Treatment Options in Neurology, 14*(2), 113-125. https://doi.org/10.1007/s11940-012-0166-9

Corey-Bloom, J., Wolfson, T., Gamst, A., Jin, S., Marcotte, T. D., Bentley, H., & Gouaux, B. (2012). Smoked cannabis for spasticity in multiple sclerosis: A randomized, placebo-controlled trial. *CMAJ : Canadian Medical Association Journal = Journal De L'Association Medicale Canadienne, 184*(10), 1143-1150. https://doi.org/10.1503/cmaj.110837

Cox-Georgian, D., Ramadoss, N., Dona, C., Basu, C. (2019). Therapeutic and Medicinal Uses of Terpenes. In: Joshee, N., Dhekney, S., Parajuli, P. (eds) Medicinal Plants. *Springer, Cham*. https://doi.org/10.1007/978-3-030-31269-5_15

Crandall, R. (2020). *Drugs and thugs: The history and future of america's war on drugs*. Yale University Press.

Cristino, L., Bisogno, T., & Di Marzo, V. (2020). Cannabinoids and the expanded endocannabinoid system in neurological disorders. *Nature Reviews Neurology, 16*(1), 9-29. https://doi.org/10.1038/s41582-019-0284-z

Crocq, M. (2020). History of cannabis and the endocannabinoid system. *Dialogues in Clinical Neuroscience, 22*(3), 223-228. https://doi.org/10.31887/DCNS.2020.22.3/mcrocq

Crow, T. J. (1972). A map of the rat mesencephalon for electrical self-stimulation. *Brain Research, 36*(2), 265-273. https://doi.org/10.1016/0006-8993(72)90734-2

Croxford, J. L. (2003). Therapeutic potential of cannabinoids in CNS disease. *CNS Drugs, 17*(3), 179-202. https://doi.org/10.2165/00023210-200317030-00004

Cuttler, C., Stueber, A., Cooper, Z. D., & Russo, E. (2024). Acute effects of cannabigerol on anxiety, stress, and mood: A double-blind, placebo-controlled, crossover, field trial. *Scientific Reports, 14*(1), 1–14. https://doi.org/10.1038/s41598-024-66879-0

Dahl, J., Ormstad, H., Aass, H. C. D., Sandvik, L., Malt, U. F., & Andreassen, O. A. (2016). Recovery from major depressive disorder episode after non-pharmacological treatment is associated with normalized cytokine levels. *Acta Psychiatrica Scandinavica, 134*(1), 40-47. https://doi.org/10.1111/acps.12576

Dalton, G. D., Bass, C. E., Van Horn, C. G., & Howlett, A. C. (2009). Signal transduction via cannabinoid receptors. *CNS & Neurological Disorders Drug Targets, 8*(6), 422-431. https://doi.org/10.2174/187152709789824615

Dash, D. K., Tyagi, C. K., Sahu, A. K., & Tripathi, V. (2022). Revisiting the medicinal value of terpenes and terpenoids. In *Revisiting Plant Biostimulants*. IntechOpen. https://doi.org/10.5772/intechopen.102612

Dasram, M. H., Naidoo, P., Walker, R. B., & Khamanga, S. M. (2024). Targeting the endocannabinoid system present in the glioblastoma tumour microenvironment as a potential anti-cancer strategy. *International Journal of Molecular Sciences, 25*(3), 1371. https://doi.org/10.3390/ijms25031371

David, M. Z., & Daum, R. S. (2010). Community-associated methicillin-resistant staphylococcus aureus: Epidemiology and clinical consequences of an emerging epidemic. *Clinical Microbiology Reviews, 23*(3), 616-687. https://doi.org/10.1128/CMR.00081-09

Dawidowicz, A. L., Olszowy-Tomczyk, M., & Typek, R. (2021). CBG, CBD, Δ9-THC, CBN, CBGA, CBDA and Δ9-THCA as antioxidant agents and their intervention abilities in antioxidant action. *Fitoterapia, 152*, 104915. https://doi.org/10.1016/j.fitote.2021.104915

Dawson D.A. (2018). Synthetic cannabinoids, organic cannabinoids, the endocannabinoid system, and their relationship to obesity, diabetes, and depression. *Mol Biol 7*: 219.

Dawson D.A. (2019a). The psychosocial and biological aspects of synthetic and natural FAAH inhibitors. *Edel J Biomed Res Rev 1*, 6-11. https://doi.org/10.33805/2690-2613.102

Dawson D.A. (2019b). Cannabidiol Psychoactivity: A perspective on claims of us patent 6630507. *Journal of Medical Biomedical and Applied Sciences.* https://doi.org/ 10.15520/jmbas.v7i1.174

Dawson, D.A. (2022). Trust decay in the American healthcare delivery system. *GSC Biological and Pharmaceutical Sciences, 20*(02), 072–079. https://doi.org/10.30574/gscbps.2022.20.2.0318

Dawson, D. A. (2024). *Structural Stigmatization of Medicinal Botanic Cannabinoid Use: An Interpretative Phenomenological Analysis* (Order No. 31484974). Available from Dissertations & Theses

@ National University; ProQuest One Academic. (3085317446). https://go.openathens.net/redirector/nu.edu?url=https://www.proquest.com/dissertations-theses/structural-stigmatization-medicinal-botanic/docview/3085317446/se-2

Dawson, D.A. & Persad, C.P. (2022). Targeting the endocannabinoid system in the treatment of addiction disorders. *GSC Biological and Pharmaceutical Sciences, 19*(02), 064–074. https://doi.org/10.30574/gscbps.2022.19.2.0175

Deger, G. E., & Quick, D. W. (2009). The enduring menace of MRSA: Incidence, treatment, and prevention in a county jail. *Journal of Correctional Health Care, 15*(3), 174-178. https://doi.org/10.1177/1078345808326623

De Gregorio, D., McLaughlin, R. J., Posa, L., Ochoa-Sanchez, R., Enns, J., Lopez-Canul, M., Aboud, M., Maione, S., Comai, S., & Gobbi, G. (2019). Cannabidiol modulates serotonergic transmission and reverses both allodynia and anxiety-like behavior in a model of neuropathic pain. *Pain, 160*(1), 136-150. https://doi.org/10.1097/j.pain.0000000000001386

Dell, D. D., & Stein, D. P. (2021). Exploring the use of medical marijuana for supportive care of oncology patients. *Journal of the Advanced Practitioner in Oncology, 12*(2), 188-201. https://doi.org/10.6004/jadpro.2021.12.2.6

de Mello Schier, A., R., de Oliveira Ribeiro, N., P., Coutinho, D. S., Machado, S., Arias-Carrión, O., Crippa, J. A., Zuardi, A. W., Nardi, A. E., & Silva, A. C. (2014). Antidepressant-like and anxiolytic-like effects of cannabidiol: A chemical compound of cannabis sativa. *CNS & Neurological Disorders Drug Targets, 13*(6), 953-960. https://doi.org/10.2174/1871527313666140612114838

De Petrocellis, L., & Di Marzo, V. (2010). Non-CB1, non-CB2 receptors for endocannabinoids, plant cannabinoids, and synthetic cannabimimetics: Focus on G-protein-coupled receptors and transient receptor potential channels. *Journal of Neuroimmune Pharmacology, 5*(1), 103–121. https://doi.org/10.1007/s11481-009-9177-z

De Petrocellis, L., Ligresti, A., Schiano Moriello, A., Iappelli, M., Verde, R., Stott, C. G., Cristino, L., Orlando, P., & Di Marzo, V. (2013). Non-THC cannabinoids inhibit prostate carcinoma growth in vitro and in vivo: pro-apoptotic effects and underlying mechanisms. *British journal of pharmacology, 168*(1), 79–102. https://doi.org/10.1111/j.1476-5381.2012.02027.x

De Petrocellis, L., Orlando, P., Moriello, S. A., Aviello, G., Stott, C., Izzo, A. A., & Di Marzo, V. (2012). Cannabinoid actions at TRPV channels: Effects on TRPV3 and TRPV4 and their potential relevance to gastrointestinal inflammation. *Acta Physiologica, 204*(2), 255-266. https://doi.org/10.1111/j.1748-1716.2011.02338.x

Detweiler, M. B., Pagadala, B., Candelario, J., Boyle, J. S., Detweiler, J. G., & Lutgens, B. W. (2016). Treatment of post-traumatic stress disorder nightmares at a veterans affairs medical center. *Journal of Clinical Medicine, 5*(12) https://doi.org/10.3390/jcm5120117

Deutsch, D. G., & Chin, S. A. (1993). Enzymatic synthesis and degradation of anandamide, a cannabinoid receptor agonist. *Biochemical Pharmacology, 46*(5), 791-796. https://doi.org/10.1016/0006-2952(93)90486-g

Devane, W. A., Hanus, L., Breuer, A., Pertwee, R. G., Stevenson, L. A., Griffin, G., Gibson, D., Mandelbaum, A., Mechoulam, R., & Etinger, A. (1992). Isolation and structure of a brain constituent that binds to the cannabinoid receptor. *Science, 258*(5090), 1946+. https://link.gale.com/apps/doc/A13320461/AONE?u=pres1571&sid=ebsco&xid=87ce1694

Diana, M., Melis, M., & Gessa, G. L. (1998). Increase in mesoprefrontal dopaminergic activity after stimulation of CB1 receptors by cannabinoids. *European Journal of Neuroscience, 10*(9), 2825-2830. https://doi.org/10.1111/j.1460-9568.1998.00292.x

Di Carlo, G., & Izzo, A. A. (2003). Cannabinoids for gastrointestinal diseases: Potential therapeutic applications. *Expert Opinion on Investigational Drugs, 12*(1), 39-49. https://doi.org/10.1517/13543784.12.1.39

Di Chiara, G., & Imperato, A. (1988). Drugs abused by humans preferentially increase synaptic dopamine concentrations in the mesolimbic system of freely moving rats. *Proceedings of the National Academy of Sciences of the United States of America, 85*(14), 5274-5278. https://doi.org/10.1073/pnas.85.14.5274

Dider, S., Ji, J., Zhao, Z., & Xie, L. (2016). Molecular mechanisms involved in the side effects of fatty acid amide hydrolase inhibitors: A structural phenomics approach to proteome-wide cellular off-target deconvolution and disease association. *NPJ Systems Biology and Applications, 2*, 16023. https://doi.org/10.1038/npjsba.2016.23

Di Filippo, M., Pini, L. A., Pelliccioli, G. P., Calabresi, P., & Sarchielli, P. (2008). Abnormalities in the cerebrospinal fluid levels of endocannabinoids in multiple sclerosis. *Journal of Neurology, Neurosurgery and Psychiatry, 79*(11), 1224.

Di Marzo, V., & Maccarrone, M. (2008). FAAH and anandamide: Is 2-AG really the odd one out? *Trends in Pharmacological Sciences, 29*(5), https://doi.org/10.1016/j.tips.2008.03.001

Di Marzo, V., Stella, N., & Zimmer, A. (2015). Endocannabinoid signalling and the deteriorating brain. *Nature Reviews Neuroscience, 16*(1), 30-42. https://doi.org/10.1038/nrn3876

Di Marzo, V., Melck, D., Bisogno, T., & De Petrocellis, L. (1998). Endocannabinoids: Endogenous cannabinoid receptor ligands with neuromodulatory action. *Trends in Neurosciences, 21*(12), 521.

Dinakar, C., & O'Connor, G.,T. (2016). The health effects of electronic cigarettes. *The New England Journal of Medicine, 375*(26), 2608–2609. https://doi.org/10.1056/NEJMc1613869

Doenni, V. M., Gray, J. M., Song, C. M., Patel, S., Hill, M. N., & Pittman, Q. J. (2016). Deficient adolescent social behavior following early-life inflammation is ameliorated by augmentation of anandamide signaling. *Brain Behavior and Immunity, 58*, 237-247. https://doi.org/10.1016/j.bbi.2016.07.152

Drobnik, J., & Drobnik, E. (2016). Timeline and bibliography of early isolations of plant metabolites (1770–1820) and their impact to pharmacy: A critical study. *Fitoterapia, 115*, 155-164. https://doi.org/10.1016/j.fitote.2016.10.009

Drug Enforcement Administration, Department of Justice. (2017). Schedules of controlled substances: Placement of FDA-approved products of oral solutions containing dronabinol [(-)-delta-9-transtetrahydrocannabinol (delta-9-THC)] in schedule II. Interim final rule, with request for comments. *Federal register, 82*(55), 14815-14820.

Drysdale, A. J., Ryan, D., Pertwee, R. G., & Platt, B. (2006). Cannabidiol-induced intracellular Ca2+ elevations in hippocampal cells. *Neuropharmacology, 50*(5), 621-631. https://doi.org/10.1016/j.neuropharm.2005.11.008

Dunlop, B. W., Kaye, J. L., Youngner, C., & Rothbaum, B. (2014). Assessing treatment-resistant posttraumatic stress disorder: The emory treatment resistance interview for PTSD (E-TRIP). *Behavioral Sciences, 4*(4), 511-527. https://doi.org/10.3390/bs4040511

Durkheim, É. (1951). *Suicide, a study in sociology*. Free Press.

Duvall, C. S. (2014). Drug laws, bioprospecting and the agricultural heritage of cannabis in africa. *Space & Polity, 20*(1), 10-25. https://doi.org/10.1080/13562576.2016.1138674

Eddin, L. B., Jha, N. K., Meeran, M. F. N., Kesari, K. K., Beiram, R., & Ojha, S. (2021). Neuroprotective potential of limonene and limonene containing natural products. *Molecules (Basel, Switzerland), 26*(15) https://doi.org/10.3390/molecules26154535

Ehlers, A., Hackmann, A., & Michael, T. (2004). Intrusive re-experiencing in post-traumatic stress disorder: Phenomenology, theory, and therapy. *Memory, 12*(4), 403-415. https://doi.org/10.1080/09658210444000025

El-Alfy, A., Ivey, K., Robinson, K., Ahmed, S., Radwan, M., Slade, D., Khan, I., ElSohly, M., & Ross, S. (2010). Antidepressant-like effect of Δ9-tetrahydrocannabinol and other cannabinoids isolated from cannabis sativa L. *Pharmacology Biochemistry and Behavior, 95*(4), 434-442. https://doi.org/10.1016/j.pbb.2010.03.004

Elliott, L., Golub, A., Bennett, A., & Guarino, H. (2015). PTSD and cannabis-related coping among recent veterans in new york city. *Contemporary Drug Problems, 42*, 60-76.

Elmer, G. I., Pieper, J. O., Hamilton, L. R., & Wise, R. A. (2010). Qualitative differences between C57BL/6J and DBA/2J mice in morphine potentiation of brain stimulation reward and intravenous self-administration. *Psychopharmacology, 208*(2), 309-321. https://doi.org/10.1007/s00213-009-1732-z

El-Remessy, A., Al-Shabrawey, M., Khalifa, Y., Tsai, N., Caldwell, R. B., & Liou, G. I. (2006). Neuroprotective and blood-retinal barrier-preserving effects of cannabidiol in experimental diabetes. *The American Journal of Pathology, 168*(1), 235-244. https://doi.org/10.2353/ajpath.2006.050500

Emboden, W. (1972). *Ritual use of Cannabis sativa L.: A historical ethnographic survey.* In Furst, P. T. (Ed.), Flesh of the Gods; the Ritual Use of Hallucinogens (pp. 214–236). New York: Praeger.

Erman, M. B., & Kane, B. J. (2008). Chemistry around pinene and pinane: A facile synthesis of cyclobutanes and oxatricyclo-derivative of pinane from cis- and trans-pinanols. *Chemistry & Biodiversity, 5*(6), 910-919. https://doi.org/10.1002/cbdv.200890104

Exum, J. J. (2019). From Warfare to Welfare: Reconceptualizing drug sentencing during the opioid crisis. *University of Kansas Law Review, 67*(5), 941–960.

Fadus, M. C., Smith, T. T., & Squeglia, L. M. (2019). The rise of e-cigarettes, pod mod devices, and JUUL among youth: Factors influencing use, health implications, and downstream effects. *Drug and Alcohol Dependence, 201*, 85–93. https://doi.org/10.1016/j.drugalcdep.2019.04.011

Fahmy, M. A., Farghaly, A. A., Hassan, E. E., Hassan, E. M., Hassan, Z. M., Mahmoud, K., & Omara, E. A. (2022). Evaluation of the anti-cancer/anti-mutagenic efficiency of lavandula officinalis essential oil. *Asian Pacific Journal of Cancer Prevention: APJCP, 23*(4), 1215-1222. https://doi.org/10.31557/APJCP.2022.23.4.1215

Fahn, S. (1989). Adverse Effects of Levodopa in Parkinson's Disease. In: Calne, D.B. (eds) Drugs for the Treatment of Parkinson's Disease. Handbook of Experimental Pharmacology, vol 88. Springer, Berlin, Heidelberg. https://doi.org/10.1007/978-3-642-73899-9_14

Fanelli, F., Di Lallo, V., D., Belluomo, I., De Iasio, R., Baccini, M., Casadio, E., Gasparini, D. I., Colavita, M., Gambineri, A., Grossi, G., Vicennati, V., Pasquali, R., & Pagotto, U. (2012). Estimation of reference intervals of five endocannabinoids and endocannabinoid related compounds in human plasma by two dimensional-LC/MS/MS. *Journal of Lipid Research, 53*(3), 481-493. https://doi.org/10.1194/jlr.M021378

Faraone, S. V., Perlis, R. H., Doyle, A. E., Smoller, J. W., Goralnick, J. J., Holmgren, M. A., & Sklar, P. (2005). Molecular genetics of attention-deficit/hyperactivity disorder. *Biological Psychiatry, 57*(11), 1313-1323. https://doi.org/10.1016/j.biopsych.2004.11.024

Fattore, L., & Fratta, W. (2011). Beyond THC: The new generation of cannabinoid designer drugs. *Frontiers in Behavioral Neuroscience, 5* https://doi.org/10.3389/fnbeh.2011.00060

Faubion, J. (2013). Reevaluating drug policy: Uruguay's efforts to reform marijuana laws. Law and Business Review of the Americas, 19(3), 383-410. https://scholar.smu.edu/lbra/vol19/iss3/7

Fellous, T., De Maio, F., Kalkan, H., Carannante, B., Boccella, S., Petrosino, S., Maione, S., Di Marzo, V., & Iannotti, F. A. (2020). Phytocannabinoids promote viability and functional adipogenesis of bone marrow-derived mesenchymal stem cells through different molecular targets. *Biochemical Pharmacology, 175*, 113859. https://doi.org/10.1016/j.bcp.2020.113859

Ferber, S. G., Namdar, D., Hen-Shoval, D., Eger, G., Koltai, H., Shoval, G., Shbiro, L., & Weller, A. (2020). The 'entourage effect': Terpenes coupled with cannabinoids for the treatment of mood disorders and anxiety disorders. *Current Neuropharmacology, 18*(2), 87-96. https://doi.org/10.2174/1570159X17666190903103923

Ferguson, S. S., & Feldman, R. D. (2014). β-adrenoceptors as molecular targets in the treatment of hypertension. *Canadian Journal of Cardiology, 30*(5), S3-S8.

Fernández-Ruiz, J., Sagredo, O., Pazos, R. M., García, C., Pertwee, R., Mechoulam, R., & Martínez-Orgado, J. (2013). Cannabidiol for neurodegenerative disorders: Important new clinical applications for this phytocannabinoid? *British Journal of Clinical Pharmacology, 75*(2), 323-333.
https://doi.org/10.1111/j.1365-2125.2012.04341.x

Fischedick, J. T. (2017). Identification of terpenoid chemotypes among high (-)-trans-Δ9-tetrahydrocannabinol-producing cannabis sativa L. cultivars. *Cannabis & Cannabinoid Research, 2*(1), 34-47.
https://doi.org/10.1089/can.2016.0040

Földy, C., Malenka, R. C., & Südhof, T.,C. (2013). Autism-associated neuroligin-3 mutations commonly disrupt tonic endocannabinoid signaling. *Neuron, 78*(3), 498-509.
https://doi.org/10.1016/j.neuron.2013.02.036

Francomano, F., Caruso, A., Barbarossa, A., Fazio, A., La Torre, C., Ceramella, J., Mallamaci, R., Saturnino, C., Iacopetta, D., & Sinicropi, M. S. (2019). B-caryophyllene: A sesquiterpene with countless biological properties. *Applied Sciences (2076-3417), 9*(24), 5420. https://doi.org/10.3390/app9245420

Frank, L. E., & Nagel, S. K. (2017). Addiction and moralization: The role of the underlying model of addiction. *Neuroethics, 10*(1), 129-139. https://doi.org/10.1007/s12152-017-9307-x

Freet, C. S., Arndt, A., & Grigson, P. S. (2013). Compared with DBA/2J mice, C57BL/6J mice demonstrate greater preference for saccharin and less avoidance of a cocaine-paired saccharin cue. *Behavioral Neuroscience, 127*(3), 474-484. https://doi.org/10.1037/a0032402

French, J. A., Koepp, M., Naegelin, Y., Vigevano, F., Auvin, S., Rho, J. M., Rosenberg, E., Devinsky, O., Olofsson, P. S., & Dichter, M. A. (2017). Clinical studies and anti-inflammatory mechanisms of treatments. *Epilepsia, 58 Suppl 3*, 69-82. https://doi.org/10.1111/epi.13779

Fridkin, S. K., Hageman, J. C., Morrison, M., Jernigan, J. A., Sanza, L. T., Como-Sabetti, K., Harriman, K., Harrison, L. H., Lynfield, R., & Farley, M. M. (2005). Methicillin-resistant staphylococcus aureus disease in three communities. *The New England Journal of Medicine, 352*(14), 1436

Fridkin, S. K., Hill, H. A., Volkova, N. V., Edwards, J. R., Lawton, R. M., Gaynes, R. P., & McGowan Jr., J. E. (2002). Temporal changes in prevalence of antimicrobial resistance in 23 U.S. hospitals. *Emerging Infectious Diseases, 8*(7), 697.

Friedman, S. R., Williams, L. D., Guarino, H., Mateu-Gelabert, P., Krawczyk, N., Hamilton, L., Walters, S. M., Ezell, J. M., Khan, M., Di Iorio, J., Yang, L. H., & Earnshaw, V. A. (2022). The stigma system: How sociopolitical domination, scapegoating, and stigma shape public health. *Journal of Community Psychology, 50*(1), 385.

Fritze, T., Fink, A., & Doblhammer, G. (2019). Dementia in an aging world. In F. Jotterand, M. Ienca, T. Wangmo & B. S. Elger (Eds.), (pp. 15-34). Oxford University Press. https://doi.org/10.1093/med/9780190459802.003.0002

Gachet, M. S., Rhyn, P., Bosch, O. G., Quednow, B. B., & Gertsch, J. ü. (2015). A quantitative LC-MS/MS method for the measurement of arachidonic acid, prostanoids, endocannabinoids, N-acylethanolamines and steroids in human plasma. *Journal of Chromatography B, 976-977*, 6-18. https://doi.org/10.1016/j.jchromb.2014.11.001

Gao, W., Walther, A., Wekenborg, M., Penz, M., & Kirschbaum, C. (2020). Determination of endocannabinoids and N-acylethanolamines in human hair with LC-MS/MS and their relation to symptoms of depression, burnout, and anxiety. *Talanta, 217* https://doi.org/10.1016/j.talanta.2020.121006

Gaoni, Y., & Mechoulam, R. (1964). Isolation, structure, and partial synthesis of an active constituent of hashish. *Journal of the American chemical society, 86*(8), 1646-1647.

Garza, M.M., (2018). They came to toil: Newspaper representations of mexicans and immigrants in the great depression. *Chiricu, 3*(1), 229. https://doi.org/10.2979/chiricu.3.1.20

Gaynor, J. S., & Muir, W. W. (2015). *Handbook of veterinary pain management*. Mosby.

Gelders, D., Patesson, R., Vandoninck, S., Steinberg, P., Van Malderen, S., Nicaise, P., De

Ruyver, B., Pelc, I., Dutta, M. J., Roe, K., & Vander Laenen, F. (2009). The influence of warning messages on the public's perception of substance use: A theoretical framework. *Government Information Quarterly, 26*(2), 349-357. https://doi.org/10.1016/j.giq.2008.11.006

Gerber, R. J. (2004). *Legalizing marijuana: Drug policy reform and prohibition politics.*

Greenwood Publishing Group.

Gessa, G. L., Melis, M., Muntoni, A. L., & Diana, M. (1998). Cannabinoids activate mesolimbic dopamine neurons by an action on cannabinoid CB1 receptors. *European Journal of Pharmacology, 341*(1), 39-44. https://doi.org/10.1016/s0014-2999(97)01442-8

Gildea, L., Ayariga, J. A., Ajayi, O. S., Xu, J., Villafane, R., & Samuel-Foo, M. (2022). Cannabis sativa CBD extract shows promising antibacterial activity against salmonella typhimurium and S. newington. *Molecules (Basel, Switzerland), 27*(9) https://doi.org/10.3390/molecules27092669

Gioé-Gallo, C., Ortigueira, S., Brea, J., Raïch, I., Azuaje, J., Paleo, M. R., Majellaro, M., Loza, M. I., Salas, C. O., García-Mera, X., Navarro, G., & Sotelo, E. (2023). Pharmacological insights emerging from the characterization of a large collection of synthetic cannabinoid receptor agonists designer drugs. *Biomedicine & Pharmacotherapy, 164* https://doi.org/10.1016/j.biopha.2023.114934

Giossi, R., Mercenari, M., Filippi, M., Zanetta, C., Antozzi, C. G., Brambilla, L., Confalonieri, P., Crisafulli, S. G., Tomas Roldan, E., Annovazzi, P., Conti, M. Z., Barrilà, C., Ronzoni, M., Grobberio, M., Negri, A., Gustavsen, S., & Torri Clerici, V. (2024). Unprescribed cannabinoids and multiple sclerosis: A multicenter, cross-sectional, epidemiological study in lombardy, italy. *Journal of Neurology,* 1-20. https://doi.org/10.1007/s00415-024-12472-4

Giuffrida, A., Parsons, L. H., Kerr, T. M., de Fonseca, F. R., Navarro, M., & Piomelli, D. (1999). Dopamine activation of endogenous cannabinoid signaling in dorsal striatum. *Nature Neuroscience,* (4) https://escholarship.org/uc/item/6p85k48q

Glaser, D. (1956). Criminality theories and behavioral images. *American Journal of Sociology, 61*(5), 433-444. http://www.jstor.org/stable/2773486

Glass, M., Dragunow, M., & Faull, R. L. (1997). Cannabinoid receptors in the human brain: A detailed anatomical and quantitative autoradiographic study in the fetal, neonatal and adult human brain. *Neuroscience, 77*(2), 299-318. https://doi.org/10.1016/s0306-4522(96)00428-9

Gold, R., & Hartung, H. (2005). Towards individualised multiple-sclerosis therapy. *The Lancet. Neurology, 4*(11), 693-694. https://doi.org/10.1016/S1474-4422(05)70205-2

Gomis-González, M., Busquets-Garcia, A., Matute, C., Maldonado, R., Mato, S., & Ozaita, A. (2016). Possible therapeutic doses of cannabinoid type 1 receptor antagonist reverses key alterations in fragile X syndrome mouse model. *Genes, 7*(9) https://doi.org/10.3390/genes7090056

Goniewicz, M. L., Knysak, J., Gawron, M., Kosmider, L., Sobczak, A., Kurek, J., Prokopowicz, A., Jablonska-Czapla, M., Rosik-Dulewska, C., Havel, C., Jacob, P. I.,II, & Benowitz, N. (2014). Levels of selected carcinogens and toxicants in vapour from electronic cigarettes. https://doi.org/10.1136/tobaccocontrol-2012-050859

Goode, E. (1993). *Drugs in american society.* 3rd Edition. New York, NY: McGraw-Hill.

Gorman, J. M. (1997). Comorbid depression and anxiety spectrum disorders. *Depression & Anxiety (1091-4269), 4*(4), 160-168. https://doi.org/10.1002/(SICI)1520-6394(1996)4:4

Gosselin, C., & Cote, G. (2001). Weight loss maintenance in women two to eleven years after participating in a commercial program: A survey. *BMC Women's Health, 1,* 2-7. https://doi.org/10.1186/1472-6874-1-2

Green, K. (1979). Marihuana in ophthalmology-past, present and future. *Annals of Ophthalmology, 11*(2), 203-205.

Griffin, I. W. C., Randall, P. K., & Middaugh, L. D. (2007). Intravenous cocaine self-administration: Individual differences in male and female C57BL/6J mice. *Pharmacology, Biochemistry and Behavior, 87*(2), 267-279. https://doi.org/10.1016/j.pbb.2007.04.023

Gururajan, A., Taylor, D. A., & Malone, D. T. (2012). Cannabidiol and clozapine reverse MK-801-induced deficits in social interaction and hyperactivity in Sprague–Dawley rats. *Journal of Psychopharmacology, 26*(10), 1317-1332. https://doi.org/10.1177/0269881112441865

Gunduz-Cinar, O., MacPherson, K. P., Cinar, R., Gamble-George, J., Sugden, K., Williams, B., Godlewski, G., Ramikie, T. S., Gorka, A. X., Alapafuja, S. O., Nikas, S. P., Makriyannis, A., Poulton, R., Patel, S., Hariri, A. R., Caspi, A., Moffitt, T. E., Kunos, G., & Holmes, A. (2013). Convergent translational evidence of a role for anandamide in amygdala-mediated fear extinction, threat processing and stress-reactivity. *Molecular Psychiatry, 18*(7), 813-823. https://doi.org/10.1038/mp.2012.72

Gurusamy, K. S., Koti, R., Toon, C. D., Wilson, P., & Davidson, B. R. (2013). Antibiotic therapy for the treatment of methicillin-resistant staphylococcus aureus (MRSA) in non surgical wounds. *The Cochrane Database of Systematic Reviews,* (11), CD010427. https://doi.org/10.1002/14651858.CD010427.pub2

Guy, R. H. (2024). Drug delivery to and through the skin. *Drug Delivery and Translational Research: An Official Journal of the Controlled Release Society,* 1-9. https://doi.org/10.1007/s13346-024-01614-w

Hall, P. (2022). *Marking the 80th anniversary of 'hemp for victory,' the government's censored cannabis film.* Accretive Capital LLC dba Benzinga.com.

Halpern, M., & Walther, S. (2010). Cannabinoids and dementia: A review of clinical and preclinical data. *Pharmaceuticals, 3*(8), 2689-2708. https://doi.org/10.3390/ph3082689

Hardiman, O., van den Berg, L. H., & Kiernan, M. C. (2011). Clinical diagnosis and management of amyotrophic lateral sclerosis. *Nature Reviews Neurology, 7*(11), 639. https://doi.org/10.1038/nrneurol.2011.153

Harris, K. N., & Martin, W. (2021). Persistent inequities in cannabis policy. *Judges' Journal, 60*(1), 9-13.

Hashemi, P., & Ahmadi, S. (2023). Alpha-pinene moderates memory impairment induced by kainic acid via improving the BDNF/TrkB/CREB signaling pathway in rat hippocampus. *Frontiers in Molecular Neuroscience, 16* https://doi.org/10.3389/fnmol.2023.1202232

Hati, D. K., Sahu, P. K., & Patro, S. K. (2024). Therapeutic uses of cannabis in indian traditional medicine with attention on gastrointestinal disorders. *Research Journal of Pharmacy and Technology, 17*(1), 229-235. https://doi.org/10.52711/0974-360X.2024.00036

Hatzenbuehler M. L. (2016). Structural stigma: Research evidence and implications for psychological science. *The American psychologist, 71*(8), 742–751. https://doi.org/10.1037/amp0000068

Hawdon, J. E. (2001). The role of presidential rhetoric in the creation of a moral panic: Reagan, Bush, and the war on drugs. *Deviant behavior, 22*(5), 419-445.

Hazekamp, A., & Fischedick, J. T. (2012). Cannabis - from cultivar to chemovar. *Drug Testing and Analysis, 4*(7-8), 660-667. https://doi.org/10.1002/dta.407

Heffler, E., Madeira, L. N. G., Ferrando, M., Puggioni, F., Racca, F., Malvezzi, L., Passalacqua, G., & Canonica, G. W. (2018). Inhaled corticosteroids safety and adverse effects in patients with asthma. *The Journal of Allergy and Clinical Immunology: In Practice, 6*(3), 776-781. https://doi.org/10.1016/j.jaip.2018.01.025

Henshaw, F. R., Dewsbury, L. S., Lim, C. K., & Steiner, G. Z. (2021). The effects of cannabinoids on pro- and anti-inflammatory cytokines: A systematic review of in vivo studies. *Cannabis & Cannabinoid Research, 6*(3), 177-195. https://doi.org/10.1089/can.2020.0105

Hepler, R. S., & Frank, I. R. (1971). Marihuana smoking and intraocular pressure. *Jama, 217*(10), 1392. https://doi.org/10.1001/jama.1971.03190100074024

Herbig, C., & Sirocko, F. (2013). Palaeobotanical evidence for agricultural activities in the eifel region during the holocene: Plant macro-remain and pollen analyses from sediments of three maar lakes in the quaternary westeifel volcanic field (germany, rheinland-pfalz). *Vegetation History and Archaeobotany: The Journal of Quaternary Plant Ecology, Palaeoclimate and Ancient Agriculture - Official Organ of the International Work Group for Palaeoethnobotany, 22*(6), 447-462. https://doi.org/10.1007/s00334-012-0387-6

Herer, J. (1993). The Emperor Wears No Clothes. Van Nuys CA.: HEMP Publishing, 1993. https://upload.wikimedia.org/wikipedia/commons/b/bd/Jack_Herer_-_The_Emperor_Wears_No_Clothes.pdf

Hergenrather, J. Y., Aviram, J., Vysotski, Y., Campisi-Pinto, S., Lewitus, G. M., & Meiri, D. (2020). Cannabinoid and terpenoid doses are associated with adult ADHD status of medical cannabis patients. *Rambam Maimonides Medical Journal, 11*(1), 1-14. https://doi.org/10.5041/RMMJ.10384

Herkenham, M., Lynn, A. B., Little, M. D., Johnson, M. R., Melvin, L. S., de Costa, B. R., & Rice, K. C. (1990). Cannabinoid receptor localization in brain. *Proceedings of the National Academy of Sciences of the United States of America, 87*(5), 1932-1936. https://doi.org/10.1073/pnas.87.5.1932

Herkenham, M., Lynn, A. B., Johnson, M. R., Melvin, L. S., de Costa, B. R., & Rice, K. C. (1991). Characterization and localization of cannabinoid receptors in rat brain: a quantitative in vitro autoradiographic study. *The Journal of neuroscience: the official journal of the Society for Neuroscience, 11*(2), 563–583.

Herndon, G. M. (1963). Hemp in colonial virginia. *Agricultural History, 37*(2), 86-93.

Hess, C., Schoeder, C. T., Pillaiyar, T., Madea, B., & Müller, C. E. (2016). Pharmacological evaluation of synthetic cannabinoids identified as constituents of spice. *Forensic Toxicology, 34*(2), 329–343. https://doi.org/10.1007/s11419-016-0320-2

Hesselink, J. M. K. (2013). Evolution in pharmacologic thinking around the natural analgesic palmitoylethanolamide: From nonspecific resistance to PPAR-α agonist and effective nutraceutical. *Journal of Pain Research, 6*, 625-634. https://doi.org/10.2147/JPR.S48653

Hill, M. N., Campolongo, P., Yehuda, R., & Patel, S. (2018). Integrating endocannabinoid signaling and cannabinoids into the biology and treatment of posttraumatic stress disorder. *Neuropsychopharmacology, 43*(1), 80-102. https://doi.org/10.1038/npp.2017.162

Hill, M. N., Miller, G. E., Ho, W. -. V., Gorzalka, B. B., & Hillard, C. J. (2008). Serum endocannabinoid content is altered in females with depressive disorders: A preliminary report. *Pharmacopsychiatry, 41*(2), 48-53. https://doi.org/10.1055/s-2007-993211

Hillen, J. B., Soulsby, N., Alderman, C., & Caughey, G. E. (2019). Safety and effectiveness of cannabinoids for the treatment of neuropsychiatric symptoms in dementia: A systematic review. *Therapeutic Advances* in *Drug Safety, 10*, 2042098619846993. https://doi.org/10.1177/2042098619846993

Hirvikoski, T., Mittendorfer-Rutz, E., Boman, M., Larsson, H., Lichtenstein, P., & Boölte, S. (2016). Premature mortality in autism spectrum disorder. *The British Journal of Psychiatry, 208*(3), 232-238.

Hirvonen, J., Goodwin, R. S., Li, C., Terry, G. E., Zoghbi, S. S., Morse, C., Pike, V. W., Volkow, N. D., Huestis, M. A., & Innis, R. B. (2012). Reversible and regionally selective downregulation of brain cannabinoid CB1 receptors in chronic daily cannabis smokers. *Molecular Psychiatry, 17*(6), 642-649. https://doi.org/10.1038/mp.2011.82

Holifield, M. C. (2013). *Blowing smoke: Harry J. Anslinger and the Marijuana Tax Act of 1937*. Arkansas State University.

Hong, M., Kim, J., Han, J., Ryu, B., Lim, Y., Lim, J., Park, S., Kim, C., Lee, S., & Kwon, T. (2023). In vitro and in vivo anti-inflammatory potential of cannabichromene isolated from hemp. *Plants, 12*(23) https://doi.org/10.3390/plants12233966

Hori, H., & Kim, Y. (2019). Inflammation and post-traumatic stress disorder. *Psychiatry and Clinical Neurosciences, 73*(4), 143-153. https://doi.org/10.1111/pcn.12820

Hoskins, M., Pearce, J., Bethell, A., Dankova, L., Barbui, C., Tol, W. A., van Ommeren, M., de Jong, J., Seedat, S., Chen, H., & Bisson, J. I. (2015). Pharmacotherapy for post-traumatic stress disorder: Systematic review and meta-analysis. *The British Journal of Psychiatry, 206*(2), 93-100. https://doi.org/10.1192/bjp.bp.114.148551

Howland, R. H. (2008). How are drugs approved? Part 1: The evolution of the food and drug administration. *Journal of Psychosocial Nursing and Mental Health Services, 46*(1), 15-19. https://doi.org/10.3928/02793695-20080101-06

Huang, Z., Zhao, Y., Li, S., Gu, H., Lin, L., Yang, Z., Niu, Y., Zhang, C., & Luo, J. (2020). Comparative efficacy and acceptability of pharmaceutical management for adults with post-traumatic stress disorder: A systematic review and meta-analysis. *Frontiers in Pharmacology, 11*, 559. https://doi.org/10.3389/fphar.2020.00559

Hurd, Y. L., Yoon, M., Manini, A. F., Hernandez, S., Olmedo, R., Ostman, M., & Jutras-Aswad, D. (2015). Early Phase in the Development of Cannabidiol as a Treatment for Addiction: Opioid Relapse Takes Initial Center Stage. *Neurotherapeutics : the journal of the American Society for Experimental NeuroTherapeutics, 12*(4), 807–815. https://doi.org/10.1007/s13311-015-0373-7

Hurgobin, B., Tamiru-Oli, M., Welling, M. T., Doblin, M. S., Bacic, A., Whelan, J., & Lewsey, M. G. (2021). Recent advances in cannabis sativa genomics research. *The New Phytologist, 230*(1), 73-89. https://doi.org/10.1111/nph.17140

Hurwitz, R., Blackmore, R., Hazell, P., Williams, K., & Woolfenden, S. (2012). Tricyclic antidepressants for autism spectrum disorders (ASD) in children and adolescents. *The Cochrane Database of Systematic Reviews,* (3), CD008372. https://doi.org/10.1002/14651858.CD008372.pub2

Ibarra-Lecue, I., Pilar-Cuéllar, F., Muguruza, C., Florensa-Zanuy, E., Díaz, Á, Urigüen, L., Castro, E., Pazos, A., & Callado, L. F. (2018). The endocannabinoid system in mental disorders: Evidence from human brain studies. *Biochemical Pharmacology, 157*, 97-107. https://doi.org/10.1016/j.bcp.2018.07.009

Ingram, G., & Pearson, O. R. (2019). Cannabis and multiple sclerosis. *Practical Neurology, 19*(4), 310-315. https://doi.org/10.1136/practneurol-2018-002137

Iseppi, R., Brighenti, V., Licata, M., Lambertini, A., Sabia, C., Messi, P., Pellati, F., & Benvenuti, S. (2019). Chemical characterization and evaluation of the antibacterial activity of essential oils from fibre-type cannabis sativa L. (hemp). *Molecules, 24*(12), 2302. https://doi.org/10.3390/molecules24122302

Janssen, W.F. (1981) The story of the laws behind the labels. *US Food and Drug Administration.* https://www.fda.gov/media/116890/download

Jbilo, O., Ravinet-Trillou, C., Arnone, M., Buisson, I., Bribes, E., Péleraux, A., Pénarier, G., Soubrié, P., Le Fur, G., Galiègue, S., & Casellas, P. (2005). The CB1 receptor antagonist rimonabant reverses the diet-induced obesity phenotype through the regulation of lipolysis and energy balance. *FASEB Journal: Official Publication of the Federation of American Societies for Experimental Biology, 19*(11), 1567-1569. https://doi.org/10.1096/fj.04-3177fje

Jia, X., Dong, G., Li, H., Brunson, K., Chen, F., Ma, M., Wang, H., An, C., & Zhang, K. (2013). The development of agriculture and its impact on cultural expansion during the late neolithic in the western loess plateau, china. *Holocene, 23*(1), 85-92. https://doi.org/10.1177/0959683612450203

Jiang, H., Wu, Y., Wang, H., Ferguson, D. K., & Li, C. (2013). Ancient plant use at the site of yuergou, xinjiang, china: Implications from desiccated and charred plant remains. *Vegetation History and Archaeobotany: The Journal of Quaternary Plant Ecology, Palaeoclimate and Ancient Agriculture - Official Organ of the International Work Group for Palaeoethnobotany, 22*(2), 129-140. https://doi.org/10.1007/s00334-012-0365-z

Jiang, W., Zhang, Y., Xiao, L., Van Cleemput, J., Ji, S. P., Bai, G., & Zhang, X. (2005). Cannabinoids promote embryonic and adult hippocampus neurogenesis and produce anxiolytic- and antidepressant-like effects. *The Journal of clinical investigation, 115*(11), 3104–3116. https://doi.org/10.1172/JCI25509

Johnson, R. (2014, February). *Hemp as an agricultural commodity*. Library of Congress Washington DC Congressional Research Service.

Jokić, S., Jerković, I., Pavić, V., Aladić, K., Molnar, M., Kovač, M. J., & Vladimir-Knežević, S. (2022). Terpenes and cannabinoids in supercritical CO_2 extracts of industrial hemp inflorescences: optimization of extraction, antiradical and antibacterial activity. *Pharmaceuticals, 15*(9), 1117. https://doi.org/10.3390/ph15091117

Julien, M. S. (1894). Chirurgie chinoise–substance anesthétique employée en Chine, dans le commencement du IIIe siècle de notre ere, pour paralysermomentanement la sensibilite. *ComptesRendus de l'Académie de Sciences, 28*, 195-198

Jung, K., Sepers, M., Henstridge, C. M., Lassalle, O., Neuhofer, D., Martin, H., Ginger, M., Frick, A., DiPatrizio, N. V., Mackie, K., Katona, I., Piomelli, D., & Manzoni, O. J. (2012). Uncoupling of the endocannabinoid signalling complex in a mouse model of fragile X syndrome. *Nature Communications,* (1) https://doi.org/10.1038/ncomms2045

Kabir, M., Stefanovski, D., Hsu, I. R., Iyer, M., Woolcott, O. O., Zheng, D., Catalano, K. J., Chiu, J. D., Kim, S. P., Harrison, L. N., Ionut, V., Lottati, M., Bergman, R. N., Richey, J. M., Kabir, M., Stefanovski, D., Hsu, I. R., Iyer, M., Woolcott, O. O., & Zheng, D. (2011). Large size cells in the visceral adipose depot predict insulin resistance in the canine model. *Obesity (19307381), 19*(11), 2121-2129. https://doi.org/10.1038/oby.2011.254

Kales, H. C., Gitlin, L. N., & Lyketsos, C. G. (2014). Detroit Expert Panel on Assessment and Management of Neuropsychiatric Symptoms of Dementia. Management of neuropsychiatric symptoms of dementia in clinical settings: recommendations from a multidisciplinary expert panel. *J Am Geriatr Soc, 62*(4), 762-769.

Karch, S. B., ElSohly, M. A., & Brenneisen, R. (2007). Chemistry and analysis of phytocannabinoids and other cannabis constituents. (pp. 17-49)

Karhson, D. S., Krasinska, K. M., Jamie, A. D., Libove, R. A., Phillips, J. M., Chien, A. S., Garner, J. P., Hardan, A. Y., & Parker, K. J. (2018). Plasma anandamide concentrations are lower in children with autism spectrum disorder. *Molecular Autism, 9*(1), 1-6. https://doi.org/10.1186/s13229-018-0203-y

Kasper, A. M., Ridpath, A. D., Gerona, R. R., Cox, R., Galli, R., Kyle, P. B., Parker, C., Arnold, J. K., Chatham-Stephens, K., Morrison, M. A., Olayinka, O., Preacely, N., Kieszak, S. M., Martin, C., Schier, J. G., Wolkin, A., Byers, P., & Dobbs, T. (2019). Severe illness associated with reported use of synthetic cannabinoids: A public health investigation (mississippi, 2015). *Clinical Toxicology*

(15563650), 57(1), 10–18. https://doi.org/10.1080/15563650.2018.1485927

Katchan, V., David, P., & Shoenfeld, Y. (2016). Cannabinoids and autoimmune diseases: A systematic review. *Autoimmunity Reviews, 15*(6), 513-528. https://doi.org/10.1016/j.autrev.2016.02.008

Katcher, B. S. (1993). Benjamin rush's educational campaign against hard drinking. *American Journal of Public Health, 83*(2), 273-281.

Katzung, B. G., & Trevor, A. J. (2020). *Basic and clinical pharmacology 15e*. McGraw-Hill Education / Medical.

Kaur, R., Sidhu, P., & Singh, S. (2016). What failed BIA 10-2474 phase I clinical trial? global speculations and recommendations for future phase I trials. *Journal of Pharmacology & Pharmacotherapeutics, 7*(3), 120-126. https://doi.org/10.4103/0976-500X.189661

Kavanagh, K. T. (2019). Control of MSSA and MRSA in the united states: Protocols, policies, risk adjustment and excuses. *Antimicrobial Resistance & Infection Control, 8*(1) https://doi.org/10.1186/s13756-019-0550-2

Kalová, H., Janečková, B., Liptáková, Z., Rosecký, J., Verner, M., Děták, M., ... & Soukupová, A. (2016). Terpenes in forest air–health benefit and healing potential. *Acta Salus Vitae, 4*(2), 61-69.

Kaplan, J. S., Stella, N., Catterall, W. A., & Westenbroek, R. E. (2017). Cannabidiol attenuates seizures and social deficits in a mouse model of dravet syndrome. *Proceedings of the National Academy of Sciences of the United States of America, 114*(42), 11229-11234.

Katsidoni, V., Anagnostou, I., & Panagis, G. (2013). Cannabidiol inhibits the reward-facilitating effect of morphine: Involvement of 5-HT1A receptors in the dorsal raphe nucleus. *Addiction Biology, 18*(2), 286-296. https://doi.org/10.1111/j.1369-1600.2012.00483.x

Kaushik, N., Negi, S., Khare, N., Mathur, R., & Jha, A. K. A Possible Role of Pulegone against Glypican-1 for the Treatment of Alzheimer's Disease through In-Silico Approach. ISSN: 2321-9653

Kenny, B. J., & Zito, P. M. (2022). Controlled substance schedules. *StatPearls,* https://www.ncbi.nlm.nih.gov/books/NBK538457/

Kerr, D. M., Downey, L., Conboy, M., Finn, D. P., & Roche, M. (2013). Alterations in the endocannabinoid system in the rat valproic acid model of autism. *Behavioural Brain Research, 249 Suppl C*, 124-132. https://doi.org/10.1016/j.bbr.2013.04.043

Kerr, D. M., Gilmartin, A., & Roche, M. (2016). Pharmacological inhibition of fatty acid amide hydrolase attenuates social behavioural deficits in male rats prenatally exposed to valproic acid. *Pharmacological Research, 113*, 228-235. https://doi.org/10.1016/j.phrs.2016.08.033

Khajuria, D. K., Karuppagounder, V., Nowak, I., Sepulveda, D. E., Lewis, G. S., Norbury, C. C., Raup-Konsavage, W., Vrana, K. E., Kamal, F., & Elbarbary, R. A. (2023). Cannabidiol and cannabigerol, nonpsychotropic cannabinoids, as analgesics that effectively manage bone fracture pain and promote healing in mice. *Journal of Bone and Mineral Research : The Official Journal of the American Society for Bone and Mineral Research, 38*(11), 1560-1576. https://doi.org/10.1002/jbmr.4902

Khurshid, H., Qureshi, I. A., Jahan, N., Went, T. R., Sultan, W., Sapkota, A., & Alfonso, M. (2021). A systematic review of fibromyalgia and recent advancements in treatment: Is medicinal cannabis a new hope? *Cureus, 13*(8), e17332. https://doi.org/10.7759/cureus.17332

Kim, D., Lee, H., Jeon, Y., Han, Y., Kee, J., Kim, H., Shin, H., Kang, J., Lee, B. S., Kim, S., Kim, S., Park, S., Choi, B., Park, S., Um, J., & Hong, S. (2015). Alpha-pinene exhibits anti-inflammatory activity through the suppression of MAPKs and the NF-κB pathway in mouse peritoneal macrophages. *The American Journal of Chinese Medicine, 43*(4), 731-742. https://doi.org/10.1142/S0192415X15500457

Kim, S. P., Woolcott, O. O., Hsu, I. R., Stefanoski, D., Nicole Harrison, L., Zheng, D., Lottati, M., Kolka, C., Catalano, K. J., Chiu, J. D., Kabir, M., Ionut, V., Bergman, R. N., & Richey, J. M. (2012). CB1 antagonism restores hepatic insulin sensitivity without normalization of adiposity in diet-induced obese dogs. *American Journal of Physiology: Endocrinology & Metabolism, 302*, E1261-E1268. https://doi.org/10.1152/ajpendo.00496.2011

Kim, S. Y., & Solomon, D. H. (2011). Comparative safety of nonsteroidal anti-inflammatory drugs. *Nature Reviews Cardiology, 8*(4), 193-195. https://doi.org/10.1038/nrcardio.2011.30

Kimmel, H. L. (2018, April). Marijuana research at the National Institute on Drug Abuse. In *2019 Institute of Cannabis Research Conference. Colorado State University-Pueblo. Library.*

Kindred, J. H., Li, K., Ketelhut, N. B., Proessl, F., Fling, B. W., Rudroff, T., Honce, J. M., & Shaffer, W. R. (2017). Cannabis use in people with parkinson's disease and multiple sclerosis: A web-based investigation. *Complementary Therapies in Medicine, 33*, 99-104. https://doi.org/10.1016/j.ctim.2017.07.002

Kirkham, T. C., Williams, C. M., Fezza, F., & Di Marzo, V. (2002). Endocannabinoid levels in rat limbic forebrain and hypothalamus in relation to fasting, feeding and satiation: Stimulation of eating by 2-arachidonoyl glycerol. *British Journal of Pharmacology, 136*(4), 550-557. https://doi.org/10.1038/sj.bjp.0704767

Klahn, P. (2020). Cannabinoids-promising antimicrobial drugs or intoxicants with benefits? *Antibiotics, 9*(6), 297. https://doi.org/10.3390/antibiotics9060297

Klein, P., Kaminski, R. M., Koepp, M., & Löscher, W. (2024). New epilepsy therapies in development. *Nature Reviews Drug Discovery*, 1-27. https://doi.org/10.1038/s41573-024-00981-w

Knörzer, K. (2000). 3000 years of agriculture in a valley of the high himalayas. *Vegetation History and Archaeobotany, 9*(4), 219-222. https://doi.org/10.1007/bf01294636

Kogan, N. M., Lavi, Y., Topping, L. M., Williams, R. O., McCann, F. E., Yekhtin, Z., Feldmann, M., Gallily, R., & Mechoulam, R. (2021). Novel CBG derivatives can reduce inflammation, pain, and obesity. *Molecules (Basel, Switzerland), 26*(18), 5601. https://doi.org/10.3390/molecules26185601

Kovač, J., Šimunović, K., Wu, Z., Klančnik, A., Bucar, F., Zhang, Q., & Možina, S. S. (2015). Antibiotic resistance modulation and modes of action of (-)-α-pinene in campylobacter jejuni. *PLoS ONE, 10*(4), 1-14. https://doi.org/10.1371/journal.pone.0122871

Krediet, E., Bostoen, T., Breeksema, J., van Schagen, A., Passie, T., & Vermetten, E. (2020). Reviewing the potential of psychedelics for the treatment of PTSD. *International Journal of Neuropsychopharmacology, 23*(6), 385. https://doi.org/10.1093/ijnp/pyaa018

Kreitzer, A. C., & Malenka, R. C. (2007). Endocannabinoid-mediated rescue of striatal LTD and motor deficits in Parkinson's disease models. *Nature, 445*(7128), 643-647. https://doi.org/10.1038/nature05506

Kristensen, P. K., Bartels, E. M., Bliddal, H., & Astrup, A. (2007). Efficacy and safety of the weight-loss drug rimonabant: A meta-analysis of randomised trials. *Lancet (London, England), 370*(9600), 1706-1713. https://doi.org/10.1016/S0140-6736(07)61721-8

Kropp, M., Golubnitschaja, O., Mazurakova, A., Koklesova, L., Sargheini, N., Vo, T. K. S., de Clerck, E., Polivka, J., Jiri, Potuznik, P., Polivka, J., Stetkarova, I., Kubatka, P., & Thumann, G. (2023). Diabetic retinopathy as the leading cause of blindness and early predictor of cascading complications—risks and mitigation. *EPMA Journal, 14*(1), 21-42. https://doi.org/10.1007/s13167-023-00314-8

Krumbholz, A., Anielski, P., Reisch, N., Schelling, G., & Thieme, D. (2013). Diagnostic value of concentration profiles of glucocorticosteroids and endocannabinoids in hair. *Therapeutic Drug Monitoring, 35*(5), 600-607. https://doi.org/10.1097/FTD.0b013e3182953e43

Krumbholz, A., Anielski, P., Reisch, N., Schelling, G., & Thieme, D. (2013). Diagnostic value of concentration profiles of glucocorticosteroids and endocannabinoids in hair. *Therapeutic Drug*

Monitoring, 35(5), 600-607.
https://doi.org/10.1097/FTD.0b013e3182953e43

Kuddus, M., Ginawi, I. A. M., & Hazimi, A. A. (2013). Cannabis sativa: An ancient wild edible plant of india. *Emirates Journal of Food and Agriculture, 25*(10), 736. https://doi.org/10.9755/ejfa.v25i10.16400

Kuhn T. S. (1962) The structure of scientific revolutions.: University of Chicago Press: Chicago.

Lah, T. T., Novak, M., Almidon, M. A. P., Marinelli, O., Barbara Žvar Baškovič, Majc, B., Mlinar, M., Roman Bošnjak, Breznik, B., Zomer, R., & Nabissi, M. (2021). Cannabigerol is a potential therapeutic agent in a novel combined therapy for glioblastoma. *Cells, 10*(340), 340. https://doi.org/10.3390/cells10020340

Lambert, D. M., & Muccioli, G. G. (2007). Endocannabinoids and related N-acylethanolamines in the control of appetite and energy metabolism: Emergence of new molecular players. *Current Opinion in Clinical Nutrition & Metabolic Care, 10*(6), 735-744.
https://doi.org/10.1097/MCO.0b013e3282f00061

Lapierre, É, Monthony, A. S., & Torkamaneh, D. (2023). Genomics-based taxonomy to clarify cannabis classification. *Genome, 66*(8), 202-211. https://doi.org/10.1139/gen-2023-0005

Laprairie, R. B., Bagher, A. M., Precious, S. V., & Denovan-Wright, E. M. (2015). Components of the endocannabinoid and dopamine systems are dysregulated in Huntington's disease: analysis of publicly available microarray datasets. *Pharmacology research & perspectives*, *3*(1), e00104. https://doi.org/10.1002/prp2.104

Laruelle, M., Abi-Dargham, A., van Dyck, C. H., Gil, R., D'Souza, C. D., Erdos, J., McCance, E., Rosenblatt, W., Fingado, C., Zoghbi, S. S., Baldwin, R. M., Seibyl, J. P., Krystal, J. H., Charney, D. S., & Innis, R. B. (1996). Single photon emission computerized tomography imaging of amphetamine-induced dopamine release in drug-free schizophrenic subjects. *Proceedings of the National Academy of Sciences of the United States of America, 93*(17), 9235–9240.
https://doi.org/10.1073/pnas.93.17.9235

Lee, E., Kang, G., & Cho, S. (2007). Effect of flavonoids on human health: Old subjects but new challenges. *Recent Patents on Biotechnology, 1*(2), 139-150. https://doi.org/10.2174/187220807780809445

Lee, B. H., Smith, T., & Paciorkowski, A. R. (2015). Autism spectrum disorder and epilepsy: Disorders with a shared biology. *Epilepsy & Behavior, 47*, 191-201. https://doi.org/10.1016/j.yebeh.2015.03.017

Lee, H., Tamia, G., Song, H., Amarakoon, D., Wei, C., & Lee, S. (2022). Cannabidiol exerts anti-proliferative activity via a cannabinoid receptor 2-dependent mechanism in human colorectal cancer cells. *International Immunopharmacology, 108*, 108865. https://doi.org/10.1016/j.intimp.2022.108865

Le Foll, B., Gorelick, D. A., & Goldberg, S. R. (2009). The future of endocannabinoid-oriented clinical research after CB1 antagonists. *Psychopharmacology, 205*(1), 171-174. https://doi.org/10.1007/s00213-009-1506-7

6+, N. (2020). US farm bills and the 'national interest': An historical research paper. *Renewable Agriculture & Food Systems, 35*(4), 358-366. https://doi.org/10.1017/S1742170518000285

Leung, J., Chan, G., Stjepanović, D., Chung, J. Y. C., Hall, W., & Hammond, D. (2022).

Prevalence and self-reported reasons of cannabis use for medical purposes in USA and canada. *Psychopharmacology, 239*(5), 1509-1519. https://doi.org/10.1007/s00213-021-06047-8

Levitt, M., Wilson, A., Bowman, D., Kemel, S., Krepart, G., Marks, V., Schipper, H., Thomson, G., Weinerman, B., & Weinerman, R. (1981). Physiologic observations in a controlled clinical trial of the antiemetic effectiveness of 5, 10, and 15 mg of delta 9-tetrahydrocannabinol in cancer chemotherapy. ophthalmologic implications. *Journal of Clinical Pharmacology, 21*, 103S-109S. https://doi.org/10.1002/j.1552-4604.1981.tb02583.x

Lewis-Bakker, M. M., Yang, Y., Vyawahare, R., & Kotra, L. P. (2019). Extractions of medical Cannabis cultivars and the role of decarboxylation in optimal receptor responses. *Cannabis and cannabinoid research*, *4*(3), 183–194. https://doi.org/10.1089/can.2018.0067

Li, H. (1973). An archaeological and historical account of cannabis in china. *Economic Botany, 28*(4), 437-448. https://doi.org/10.1007/bf02862859

Lindan, S., Throckmorton D. (2019). FDA role in regulation of cannabis products: NIDA senior advisor for the Policy Office of Foods and Veterinary Medicine. [Internet]. Available from download (fda.gov)

Liu, Y., Wang, Y., & Jiang, C. (2017). Inflammation: The common pathway of stress-related diseases. *Frontiers in Human Neuroscience, 11*

Loflin, M., & Earleywine, M. (2015). No smoke, no fire: What the initial literature suggests regarding vapourized cannabis and respiratory risk. *Canadian Journal of Respiratory Therapy, 51*(1), 7-9

Lopez, G. (2020). Trump's criminal justice policy, explained. *Vox* Retrieved from https://www.msn.com/en-us/news/politics/trumps-criminal-justice-policy-explained/ar- B18R8nt

Lovato G. (2009). Prevent the spread of MRSA. *Materials management in health care*, *18*(8), 26–28.

Lu, H., Wang, Q., Jiang, X., Zhao, Y., He, M., & Wei, M. (2023). The potential mechanism of cannabidiol (CBD) treatment of epilepsy in pentetrazol (PTZ) kindling mice uncovered by multi-omics analysis. *Molecules, 28*(6), 2805. https://doi.org/10.3390/molecules28062805

Luchicchi, A., Lecca, S., Carta, S., Pillolla, G., Muntoni, A. L., Yasar, S., Goldberg, S. R., & Pistis, M. (2010). Effects of fatty acid amide hydrolase inhibition on neuronal responses to nicotine, cocaine and morphine in the nucleus accumbens shell and ventral tegmental area: Involvement of PPAR-α nuclear receptors. *Addiction Biology, 15*(3), 277-288. https://doi.org/10.1111/j.1369-1600.2010.00222.x

Luginbuhl, A. M. (2001). Industrial hemp (*Cannabis Savita L*): The geography of a controversial plant. *California Geographer, 41*, 1-14.

Luján, M. Á, Castro-Zavala, A., Alegre-Zurano, L., & Valverde, O. (2018). Repeated cannabidiol treatment reduces cocaine intake and modulates neural proliferation and CB1R expression in the mouse hippocampus. *Neuropharmacology, 143*, 163-175. https://doi.org/10.1016/j.neuropharm.2018.09.043

Lustig, R. H. (2017). *The hacking of the american mind: The science behind the corporate takeover of our bodies and brains*. Avery.

MacCallum, C. A., & Russo, E. B. (2018). Practical considerations in medical cannabis administration and dosing. *European Journal of Internal Medicine, 49*, 12-19. https://doi.org/10.1016/j.ejim.2018.01.004

Maccarrone, M. (2022). Tribute to professor raphael mechoulam, the founder of cannabinoid and endocannabinoid research. *Molecules, 27*(1), 323. https://doi.org/10.3390/molecules27010323

MacDonald, E., & Adams, A. (2019). *The Use of Medical Cannabis with Other Medications: A Review of Safety and Guidelines - An Update*. Canadian Agency for Drugs and Technologies in Health.

MacGillivray N. (2017). Sir William Brooke O'Shaughnessy (1808-1889), MD, FRS, LRCS Ed: Chemical pathologist, pharmacologist and pioneer in electric telegraphy. *Journal of medical biography, 25*(3), 186–196. https://doi.org/10.1177/0967772015596276

McNabb, M., Mandile, S., Ritter, D. J., Brum, A., Bacon, R., MacCaffrie, R., ... & Tangney, E. (2020). The 2019 Veterans Health and Medical Cannabis Study. *Cannabis Patient Care, 1*(1), 6-18.

Machado, K. D. C., Paz, M. F. C. J., Oliveira Santos, J. V. D., da Silva, F. C. C., Tchekalarova, J. D., Salehi, B., ... & Cavalcante, A. A. D. C. M. (2020). Anxiety therapeutic interventions of β-caryophyllene: A laboratory-based study. *Natural Product Communications, 15*(10), https://doi.org/10.1177/1934578X20962229

McKallip, R. J., Lombard, C., Fisher, M., Martin, B. R., Ryu, S., Grant, S., Nagarkatti, P. S., & Nagarkatti, M. (2002). Targeting CB2 cannabinoid receptors as a novel therapy to treat malignant lymphoblastic disease. *Blood, 100*(2), 627-634. https://doi.org/10.1182/blood-2002-01-0098

McLaughlin, R. J., & Gobbi, G. (2012). Cannabinoids and emotionality: A neuroanatomical perspective. *Neuroscience, 204*, 134-144. https://doi.org/10.1016/j.neuroscience.2011.07.052

McMahon, A. N., Varma, D. S., Fechtel, H., Sibille, K., Li, Z., Cook, R. L., & Wang, Y. (2023). Perceived effectiveness of medical cannabis among adults with chronic pain: Findings from interview data in a three-month pilot study. *Cannabis (Albuquerque, N.M.), 6*(2), 62-75. https://doi.org/10.26828/cannabis/2023/000149

McPartland, J.M. (2017). *Cannabis sativa* and *Cannabis indica* versus "Sativa" and "Indica". In: Chandra, S., Lata, H., ElSohly, M. (eds) Cannabis sativa L. - Botany and Biotechnology. Springer, Cham. https://doi.org/10.1007/978-3-319-54564-6_4

McPartland, J. M., Duncan, M., Di Marzo, V., & Pertwee, R. G. (2015). Are cannabidiol and Δ (9) -tetrahydrocannabivarin negative modulators of the endocannabinoid system? A systematic review. *British Journal of Pharmacology, 172*(3), 737-753. https://doi.org/10.1111/bph.12944

McPartland, J. M., & Guy, G. W. (2017). Models of cannabis taxonomy, cultural bias, and conflicts between scientific and vernacular names. *Botanical Review, 83*(4), 327–381. https://doi.org/10.2307/45212037

McWilliams J. C. (1989). Unsung partner against crime: Harry J. Anslinger and the Federal Bureau of Narcotics, 1930-1962. *The Pennsylvania magazine of history and biography, 113*(2), 207–236.

Magnusson, B. M., Walters, K. A., & Roberts, M. S. (2001). Veterinary drug delivery: Potential for skin penetration enhancement. *Advanced Drug Delivery Reviews, 50*(3), 205-227. https://doi.org/10.1016/S0169-409X(01)00158-2

Mahdizadeh, S., Khaleghi Ghadiri, M., & Gorji, A. (2015). Avicenna's Canon of Medicine: a review of analgesics and anti-inflammatory substances. *Avicenna journal of phytomedicine, 5*(3), 182–202.

Mahmud, M. S., Hossain, M. S., ATMF, A., Islam, M. Z., Sarker, M. E., & Islam, M. R. (2021). Antimicrobial and antiviral (SARS-CoV-2) potential of cannabinoids and cannabis sativa: A comprehensive review. *Molecules (Basel, Switzerland), 26*(23) https://doi.org/10.3390/molecules26237216

Maldonado, R., Valverde, O., & Berrendero, F. (2006). Involvement of the endocannabinoid system in drug addiction. *Trends in Neurosciences, 29*(4), 225.

Mallet, C., Dubray, C., & Dualé, C. (2016). FAAH inhibitors in the limelight, but regrettably. *International Journal of Clinical Pharmacology and Therapeutics, 54*(7), 498-501. https://doi.org/10.5414/CP202687

Malone, D. T., & Taylor, D. A. (1999). Modulation by fluoxetine of striatal dopamine release following Delta9-tetrahydrocannabinol: a microdialysis study in conscious rats. *British journal of pharmacology, 128*(1), 21–26. https://doi.org/10.1038/sj.bjp.0702753

Mannion, A., & Leader, G. (2013). Comorbidity in autism spectrum disorder: A literature review. *Research in Autism Spectrum Disorders, 7*(12), 1595-1616. https://doi.org/10.1016/j.rasd.2013.09.006

Marco, E. M., & Laviola, G. (2012). The endocannabinoid system in the regulation of emotions throughout lifespan: A discussion on therapeutic perspectives. *Journal of Psychopharmacology, 26*(1), 150-163. https://doi.org/10.1177/0269881111408459

María-Ríos, C. E., & Morrow, J. D. (2020). Mechanisms of shared vulnerability to post-traumatic stress disorder and substance use disorders. *Frontiers in Behavioral Neuroscience, 14* https://doi.org/10.3389/fnbeh.2020.00006

Marsicano, G., & Lutz, B. (2006). Neuromodulatory functions of the endocannabinoid system. *Journal of Endocrinological Investigation, 29*(3), 27-46.

Martín-Sánchez, E., Furukawa, T. A., Taylor, J., & Martin, J. L. R. (2009). Systematic review and meta-analysis of cannabis treatment for chronic pain. *Pain Medicine, 10*(8), 1353-1368. https://doi.org/10.1111/j.1526-4637.2009.00703.x

Martínez, V., Iriondo De-Hond, A., Borrelli, F., Capasso, R., Del Castillo, M. D., & Abalo, R. (2020). Cannabidiol and other non-psychoactive cannabinoids for prevention and treatment of gastrointestinal disorders: Useful nutraceuticals? *International Journal of Molecular Sciences, 21*(9) https://doi.org/10.3390/ijms21093067

Martín-Sánchez, E., Furukawa, T. A., Taylor, J., & Martin, J. L. R. (2009). Systematic review and meta-analysis of cannabis treatment for chronic pain. *Pain Medicine, 10*(8), 1353-1368. https://doi.org/10.1111/j.1526-4637.2009.00703.x

Martin-Santos, R., Crippa, J. A., Batalla, A., Bhattacharyya, S., Atakan, Z., Borgwardt, S., Allen, P., Seal, M., Langohr, K., Farré, M., Zuardi, A. W., & McGuire, P. K. (2012). Acute effects of a single, oral dose of d9-tetrahydrocannabinol (THC) and cannabidiol (CBD) administration in healthy volunteers. *Current Pharmaceutical Design, 18*(32), 4966-4979. https://doi.org/10.2174/138161212802884780

Maslow, A. H. (2017). *A theory of human motivation.* Dancing Unicorn Books.

Mayo, L. M., Rabinak, C. A., Hill, M. N., & Heilig, M. (2022). Targeting the endocannabinoid system in the treatment of posttraumatic stress disorder: A promising case of preclinical-clinical translation? *Biological Psychiatry, 91*(3), 262-272. https://doi.org/10.1016/j.biopsych.2021.07.019

Mayorga Anaya, H. J., Torres Ortiz, M. P., Flórez Valencia, D. H., & Gomezese Ribero, O. F. (2021). Efficacy of cannabinoids in fibromyalgia: A literature review. *Colombian Journal of Anesthesiology*

/ *Revista Colombiana De Anestesiología, 49*(4), 1-13. https://doi.org/10.5554/22562087.e980

Mechoulam, R., Ben-Shabat, S., Hanus, L., Ligumsky, M., Kaminski, N. E., Schatz, A. R., Gopher, A., Almog, S., Martin, B. R., Compton, D. R., & et. al. (1995). Identification of an endogenous 2-monoglyceride, present in canine gut, that binds to cannabinoid receptors. *Biochemical Pharmacology, 50*(1), 83–90. https://doi.org/10.1016/0006-2952(95)00109-d

Mechoulam, R., & Gaoni, Y. (1967). Recent advances in the chemistry of hashish. *Fortschritte der Chemie organischer Naturstoffe = Progress in the chemistry of organic natural products. Progres dans la chimie des substances organiques naturelles, 25*, 175–213. https://doi.org/10.1007/978-3-7091-8164-5_6

Melancia, F., Schiavi, S., Servadio, M., Cartocci, V., Campolongo, P., Palmery, M., Pallottini, V., & Trezza, V. (2018). Sex-specific autistic endophenotypes induced by prenatal exposure to valproic acid involve anandamide signalling. *British Journal of Pharmacology, 175*(18), 3699-3712. https://doi.org/10.1111/bph.14435

Melis, M., & Pistis, M. (2012). Hub and switches: Endocannabinoid signalling in midbrain dopamine neurons. *Philosophical Transactions: Biological Sciences, 367*(1607), 3276-3285.

Moloney, M. G. (2016). Natural products as a source for novel antibiotics. *Trends in Pharmacological Sciences, 37*(8), 689. https://doi.org/10.1016/j.tips.2016.05.001

Monthony, A. S., Page, S. R., Hesami, M., & Andrew Maxwell P. Jones. (2021). The past, present and future of cannabis sativa tissue culture. *Plants, 10*(185), 185. https://doi.org/10.3390/plants10010185

Moraes, L. J., Miranda, M. B., Loures, L. F., Mainieri, A. G., & Mármora, C. H. C. (2018). A systematic review of psychoneuroimmunology-based interventions. *Psychology, Health & Medicine, 23*(6), 635-652. https://doi.org/10.1080/13548506.2017.1417607

Moreira, F. A., Kaiser, N., Monory, K., & Lutz, B. (2008). Reduced anxiety-like behaviour induced by genetic and pharmacological inhibition of the endocannabinoid-degrading enzyme fatty acid amide hydrolase (FAAH) is mediated by CB1 receptors. *Neuropharmacology, 54*(1), 141. https://doi.org/10.1016/j.neuropharm.2007.07.005

Morelli, M. S., & O'Brien, F. X. (2001). Stevens-johnson syndrome and cholestatic hepatitis. *Digestive Diseases and Sciences, 46*(11), 2385-2388. https://doi.org/10.1023/A:1012351231143

Morena, M., Patel, S., Bains, J. S., & Hill, M. N. (2016). Neurobiological interactions between stress and the endocannabinoid system. *Neuropsychopharmacology, 41*(1), 80-102. https://doi.org/10.1038/npp.2015.166

Morgan, C. J., Das, R. K., Joye, A., Curran, H. V., & Kamboj, S. K. (2013). Cannabidiol reduces cigarette consumption in tobacco smokers: preliminary findings. *Addictive behaviors, 38*(9), 2433–2436. https://doi.org/10.1016/j.addbeh.2013.03.011

Mudge, E. M., Murch, S. J., & Brown, P. N. (2018). Chemometric analysis of cannabinoids: Chemotaxonomy and domestication syndrome. *Scientific Reports, 8*(1) https://doi.org/10.1038/s41598-018-31120-2

Muhle, R., Trentacoste, S. V., & Rapin, I. (2004). The genetics of autism. *Pediatrics, 113*(5), e472-e486.

Mukherjee, P. K., Banerjee, S., & Kar, A. (2021). Molecular combination networks in medicinal plants: Understanding synergy by network pharmacology in indian traditional medicine. *Phytochemistry Reviews: Fundamentals and Perspectives of Natural Products Research, 20*(4), 693-703. https://doi.org/10.1007/s11101-020-09730-4

Mukherjee, A., von Brömssen, M., Scanlon, B. R., Bhattacharya, P., Fryar, A. E., Hasan, M. A., Ahmed, K. M., Chatterjee, D., Jacks, G., & Sracek, O. (2008). Hydrogeochemical comparison and effects of overlapping redox zones on groundwater arsenic near the western (bhagirathi sub-basin, india) and eastern (meghna sub-basin, bangladesh) margins of the bengal basin. *Journal of Contaminant Hydrology, 99*(1-4), 31-48. https://doi.org/10.1016/j.jconhyd.2007.10.005

Mulvihill, J. J., Cunnane, E. M., Ross, A. M., Duskey, J. T., Tosi, G., & Grabrucker, A. M. (2020). Drug delivery across the blood-brain barrier: Recent advances in the use of nanocarriers. *Nanomedicine (London, England), 15*(2), 205-214. https://doi.org/10.2217/nnm-2019-0367

Murase, R., Kawamura, R., Singer, E., Pakdel, A., Sarma, P., Judkins, J., Elwakeel, E., Dayal, S., Martinez - Martinez, E., Amere, M., Gujjar, R., Mahadevan, A., Desprez, P. - Y., & McAllister, S. D. (2014). Targeting multiple cannabinoid anti-tumour pathways with a resorcinol derivative leads to inhibition of advanced stages of breast cancer. *British Journal of Pharmacology, 171*(19), 4464-4477. https://doi.org/10.1111/bph.12803

Murataeva, N., Dhopeshwarkar, A., Yin, D., Mitjavila, J., Bradshaw, H., Straiker, A., & Mackie, K. (2016). Where's my entourage? the curious case of 2-oleoylglycerol, 2-linolenoylglycerol, and 2-palmitoylglycerol. *Pharmacological Research, 110*, 173-180. https://doi.org/10.1016/j.phrs.2016.04.015

Murataeva, N., Straiker, A., & Mackie, K. (2014). Parsing the players: 2-arachidonoylglycerol synthesis and degradation in the CNS. *British Journal of Pharmacology, 171*(6), 1379-1391. https://doi.org/10.1111/bph.12411

Murphy, T. M., Ben-Yehuda, N., Taylor, R. E., & Southon, J. R. (2011). Hemp in ancient rope and fabric from the christmas cave in israel: Talmudic background and DNA sequence identification. *Journal of Archaeological Science, 38*(10), 2579-2588. https://doi.org/10.1016/j.jas.2011.05.004

Murray, R. M., Quigley, H., Quattrone, D., Englund, A., & Di Forti, M. (2016). Traditional marijuana, high-potency cannabis and synthetic cannabinoids: Increasing risk for psychosis. *World Psychiatry: Official Journal of the World Psychiatric Association (WPA), 15*(3), 195–204. https://doi.org/10.1002/wps.20341

Musto, D. F. (1999). *The american disease. [electronic resource]: Origins of narcotic control* (3rd ed. ed.). Oxford University Press.

Musto, D. F., & Korsmeyer, P. (2002). *The quest for drug control. [electronic resource]: Politics and federal policy in a period of increasing substance abuse, 1963-1981.* Yale University Press.

Mwanza, C., Chen, Z., Zhang, Q., Chen, S., Wang, W., & Deng, H. (2016). Simultaneous HPLC-APCI-MS/MS quantification of endogenous cannabinoids and glucocorticoids in hair. *Journal of Chromatography B, 1028*, 1-10. https://doi.org/10.1016/j.jchromb.2016.06.002

Nachnani, R., Raup-Konsavage, W., & Vrana, K. E. (2021). The pharmacological case for cannabigerol. *The Journal of Pharmacology and Experimental Therapeutics, 376*(2), 204-212. https://doi.org/10.1124/jpet.120.000340

Nadal, X., Del Río, C., Casano, S., Palomares, B., Ferreiro-Vera, C., Navarrete, C., Sánchez-Carnerero, C., Cantarero, I., Bellido, M. L., Meyer, S., Morello, G., Appendino, G., & Muñoz, E. (2017). Tetrahydrocannabinolic acid is a potent PPARγ agonist with neuroprotective activity. *British Journal of Pharmacology, 174*(23), 4263-4276. https://doi.org/10.1111/bph.14019

Nadir, I., Rana, N. F., Ahmad, N. M., Tanweer, T., Batool, A., Taimoor, Z., Riaz, S., & Ali, S. M. (2020). Cannabinoids and terpenes as an antibacterial and antibiofouling promotor for PES water filtration membranes. *Molecules, 25*(3), 691. https://doi.org/10.3390/molecules25030691

Nahler, G. (2022). Cannabidiol and other phytocannabinoids as cancer therapeutics. *Pharmaceutical Medicine, 36*(2), 99-129. https://doi.org/10.1007/s40290-022-00420-4

Nahtigal, I., Blake, A., Hand, A., Florentinus-Mefailoski, A., Hashemi, H., & Friedberg, J. (2016). The pharmacological properties of cannabis. *Journal of Pain Management, 9*(4), 481-491. In Pain Management Yearbook 2016; Merrick, J., Ed.; Nova Science Publishers, Inc.: New York, NY, USA, 2016; ISBN 978-1-53610-949-8.

Narayan, A. J., Downey, L. A., Manning, B., & Hayley, A. C. (2022). Cannabinoid treatments for anxiety: A systematic review and consideration of the impact of sleep disturbance. *Neuroscience and Biobehavioral Reviews, 143* https://doi.org/10.1016/j.neubiorev.2022.104941

National Academies of Sciences, Engineering, and Medicine, Health and, M. D., Board on Population Health and Public, Health Practice, & Committee on the Health Effects of Marijuana: An Evidence Review and, Research Agenda. (2017). *The health effects of cannabis and cannabinoids: The current state of evidence and recommendations for research.* National Academies Press.

National Institutes of Health. (2019). Who We Are [Internet]. *National Institutes of Health (NIH).* Available from: https://www.nih.gov/about-nih/who-we-are

Neely, A. N., & Maley, M. P. (2000). Survival of enterococci and staphylococci on hospital fabrics and plastic. *Journal of Clinical Microbiology, 38*(2), 724-726. https://doi.org/10.1128/JCM.38.2.724-726.2000

Nevalainen, T., & Irving, A. J. (2010). GPR55, a lysophosphatidylinositol receptor with cannabinoid sensitivity? *Current Topics in Medicinal Chemistry, 10*(8), 799-813. https://doi.org/10.2174/156802610791164229

Ney, L. J., Matthews, A., Bruno, R., & Felmingham, K. L. (2019). Cannabinoid interventions for PTSD: Where to next? *Progress in Neuropsychopharmacology & Biological Psychiatry, 93*, 124-140. https://doi.org/10.1016/j.pnpbp.2019.03.017

Ney, L. J., Cooper, J., Lam, G. N., Moffitt, K., Nichols, D. S., Mayo, L. M., & Lipp, O. V. (2023). Hair endocannabinoids predict physiological fear conditioning and salivary endocannabinoids predict subjective stress reactivity in humans. *Psychoneuroendocrinology, 154* https://doi.org/10.1016/j.psyneuen.2023.106296

Nguyen, N. T., Nguyen, X. T., Lane, J., & Wang, P. (2011). Relationship between obesity and diabetes in a US adult population: Findings from the national health and nutrition examination survey, 1999–2006. *Obesity Surgery,* (3) https://doi.org/10.1007/s11695-010-0335-4

Nicholas, P., & Churchill, A. (2012). The federal bureau of narcotics, the states, and the origins of modern drug enforcement in the united states, 1950-1962. *Contemporary Drug Problems, 39*(4), 595-640.

NIDA Overdose Deaths. (2020). [Internet] Available from Overdose Death Rates | National Institute on Drug Abuse (NIDA) https://nida.nih.gov/research-topics/trends-statistics/overdose-death-rates

NIDA (2020b). How effective is drug addiction treatment? [Internet] Available from https://www.drugabuse.gov/publications/principles-drug-addiction-treatment-research-based-guide-third-edition/frequently-asked-questions/how-effective-drug-addiction-treatment on 2020

NIDA. 2021, April 13. How does marijuana produce its effects? Available from https://nida.nih.gov/publications/research-reports/marijuana/how-does-marijuana-produce-its-effects

Nielsen, S., Murnion, B., Campbell, G., Young, H., & Hall, W. (2019). Cannabinoids for the treatment of spasticity. *Developmental Medicine & Child Neurology, 61*(6), 631-638. https://doi.org/10.1111/dmcn.14165

Noriega, P. (2019). *Terpenes in essential oils: Bioactivity and applications*

North, C. S., Surís, A. M., Smith, R. P., & King, R. V. (2016). The evolution of PTSD criteria across editions of DSM. *, 28, 3, 28*(3), 197-208.

Nutt, D. J., Lingford-Hughes, A., Erritzoe, D., & Stokes, P. R. A. (2015). The dopamine theory of addiction: 40 years of highs and lows. *Nature Reviews Neuroscience, 16*(5), 305. https://doi.org/10.1038/nrn3939

Nuutinen, T. (2018). Medicinal properties of terpenes found in cannabis sativa and humulus lupulus. *European Journal of Medicinal Chemistry, 157,* 198-228. https://doi.org/10.1016/j.ejmech.2018.07.076

Oakes, M., Law, W. J., & Komuniecki, R. (2019). Cannabinoids stimulate the TRP channel-dependent release of both serotonin and dopamine to modulate behavior in C elegans. *The Journal of Neuroscience, 39*(21), 4142-4152. https://doi.org/10.1523/JNEUROSCI.2371-18.2019

Okusanya, B. O., Lott, B. E., Ehiri, J., McClelland, J., & Rosales, C. (2022). Medical cannabis for the treatment of migraine in adults: A review of the evidence. *Frontiers in Neurology, 13*, 871187. https://doi.org/10.3389/fneur.2022.871187

Onaivi, E. S., Singh Chauhan, B. P., & Sharma, V. (2020). Challenges of cannabinoid delivery: How can nanomedicine help? *Nanomedicine (London, England), 15*(21), 2023-2028. https://doi.org/10.2217/nnm-2020-0221

Ong, A. D., Benson, L., Zautra, A. J., & Ram, N. (2018). Emodiversity and biomarkers of inflammation. *Emotion, 18*(1), 3-14. https://doi.org/10.1037/emo0000343

Orsolini, L., Chiappini, S., Volpe, U., Berardis, D. D., Latini, R., Papanti, G. D., & Corkery, A. J. M. (2019). Use of medicinal cannabis and synthetic cannabinoids in post-traumatic stress disorder (PTSD): A systematic review. *Medicina (Kaunas, Lithuania), 55*(9) https://doi.org/10.3390/medicina55090525

Ortiz, N. R., & Preuss, C. V. (2022). Controlled substance act. *StatPearls*, https://www.ncbi.nlm.nih.gov/books/NBK574544/

O'Shaughnessy, W. B. (1839). On the preparations of the Indian Hemp, or Gunjah. *The British and Foreign Medical Review, 10*(19), 225-228.

O'Shaughnessy W. B. (1843). On the preparations of the indian hemp, or gunjah: Cannabis indica their effects on the animal system in health, and their utility in the treatment of tetanus and other convulsive diseases. *Provincial Medical Journal and Retrospect of the Medical Sciences, 5*(123), 363–369. https://www.ncbi.nlm.nih.gov/pmc/articles/PMC2490264/pdf/provmedsurgj00865-0001.pdf

Ottria, R., Ravelli, A., Gigli, F., & Ciuffreda, P. (2014). Simultaneous ultra-high performance liquid chromatograpy-electrospray ionization-quadrupole-time of flight mass spectrometry quantification of endogenous anandamide and related N-acylethanolamides in biomatrices. *Journal of Chromatography B, 958*, 83-89. https://doi.org/10.1016/j.jchromb.2014.03.019

Oz, M., Jaligam, V., Galadari, S., Petroianu, G., Shuba, Y. M., & Shippenberg, T. S. (2010). The endogenous cannabinoid, anandamide, inhibits dopamine transporter function by a receptor-independent mechanism. *Journal of Neurochemistry, 112*(6), 1454-1464. https://doi.org/10.1111/j.1471-4159.2009.06557.x

Ożarowski, M., Karpiński, T., M., Zielińska, A., Souto, E. B., & Wielgus, K. (2021).

Cannabidiol in neurological and neoplastic diseases: Latest developments on the molecular mechanism of action. *International Journal of Molecular Sciences, 22*(9) https://doi.org/10.3390/ijms22094294

Pace, T. W. W., & Heim, C. M. (2011). A short review on the psychoneuroimmunology of posttraumatic stress disorder: From risk factors to medical comorbidities. *Brain Behavior and Immunity, 25*(1), 6-13. https://doi.org/10.1016/j.bbi.2010.10.003

Pacher, P., Bátkai, S., & Kunos, G. (2006). The endocannabinoid system as an emerging target of pharmacotherapy. *Pharmacological reviews, 58*(3), 389–462. https://doi.org/10.1124/pr.58.3.2

Paes-Colli, Y., Aguiar, A. F. L., Isaac, A. R., Ferreira, B. K., Campos, R. M. P., Trindade, P. M.

P., de Melo Reis, R. A., & Sampaio, L. S. (2022). Phytocannabinoids and cannabis-based products as alternative pharmacotherapy in neurodegenerative diseases: From hypothesis to clinical practice. *Frontiers in Cellular Neuroscience, 16*, 917164. https://doi.org/10.3389/fncel.2022.917164

Pamplona, F. A., da Silva, L. R., & Coan, A. C. (2019). Potential clinical benefits of CBD-rich cannabis extracts over purified CBD in treatment-resistant epilepsy: Observational data meta-analysis. *Frontiers in Neurology, 9*, 759. https://doi.org/10.3389/fneur.2018.01050

Pankow, J. F., Strongin, R. M., & Peyton, D. H. (2015). More on hidden formaldehyde in E-cigarette aerosols. *The New England Journal of*

Medicine, 372(16), 1576–1577 https://doi.org/10.1056/NEJMc1502242

Parker, E. D., Lin, J., Mahoney, T., Ume, N., Yang, G., Gabbay, R. A., ElSayed, N. A., & Bannuru, R. R. (2024). Economic costs of diabetes in the U.S. in 2022. *Diabetes Care, 47*(1), 26-43. https://doi.org/10.2337/dci23-0085

Parsons, L. H., & Hurd, Y. L. (2015). Endocannabinoid signalling in reward and addiction. *Nature reviews. Neuroscience, 16*(10), 579–594. https://doi.org/10.1038/nrn4004

Pastore, M. N., Kalia, Y. N., Horstmann, M., & Roberts, M. S. (2015). Transdermal patches: History, development and pharmacology. *British Journal of Pharmacology, 172*(9), 2179-2209. https://doi.org/10.1111/bph.13059

Patel, A. V., & Shah, B. N. (2018). Transdermal drug delivery system: A review. *Pharma Science Monitor, 9*(1), 378-390.

Patel, S., Rademacher, D. J., & Hillard, C. J. (2003). Differential regulation of the endocannabinoids anandamide and 2-arachidonylglycerol within the limbic forebrain by dopamine receptor activity. *The Journal of pharmacology and experimental therapeutics, 306*(3), 880–888. https://doi.org/10.1124/jpet.103.054270

Patel, S., & Hillard, C. J. (2009). Role of endocannabinoid signaling in anxiety and depression. *Current topics in behavioral neurosciences, 1*, 347–371. https://doi.org/10.1007/978-3-540-88955-7_14

Patton, D. V. (2020). A history of united states cannabis law. *Journal of Law and Health, 34*(1), 1-30.

Pautex, S., Bianchi, F., Daali, Y., Augsburger, M., de Saussure, C., Wampfler, J., Curtin, F., Desmeules, J., & Broers, B. (2022). Cannabinoids for behavioral symptoms in severe dementia: Safety and feasibility in a long-term pilot observational study in nineteen patients. *Frontiers in Aging Neuroscience, 14* https://doi.org/10.3389/fnagi.2022.957665

Peng, J., Fan, M., An, C., Ni, F., Huang, W., & Luo, J. (2022). A narrative review of molecular mechanism and therapeutic effect of cannabidiol (CBD). *Basic & Clinical Pharmacology & Toxicology, 130*(4), 439-456. https://doi.org/10.1111/bcpt.13710

Pepper, I., Vinik, A., Lattanzio, F., McPheat, W., & Dobrian, A. (2019). Countering the modern metabolic disease rampage with ancestral endocannabinoid system alignment. *Frontiers in Endocrinology, 10*, 311. https://doi.org/10.3389/fendo.2019.00311

Peprah, K., & McCormack, S. (2019). Medical cannabis for the treatment of dementia: A review of clinical effectiveness and guidelines. https://www.ncbi.nlm.nih.gov/books/NBK546328/

Pertwee, R. G. (2005). Pharmacological actions of cannabinoids. *Handbook of Experimental Pharmacology,* (168), 1-51. https://doi.org/10.1007/3-540-26573-2_1

Pertwee, R. G. (2014). Elevating endocannabinoid levels: Pharmacological strategies and potential therapeutic applications. *The Proceedings of the Nutrition Society, 73*(1), 96-105. https://doi.org/10.1017/S0029665113003649

Pescosolido, B. A., Martin, J. K., Long, J. S., Medina, T. R., Phelan, J. C., & Link, B. G. (2010). 'A disease like any other'? A decade of change in public reactions to schizophrenia, depression, and alcohol dependence. *American Psychiatric Association. 167*(11), 1321-1330. https://doi.org/10.1176/appi.ajp.2010.09121743

Petrosino, S., & Di Marzo, V. (2017). The pharmacology of palmitoylethanolamide and first data on the therapeutic efficacy of some of its new formulations. *British Journal of Pharmacology, 174*(11), 1349-1365. https://doi.org/10.1111/bph.13580

Phan, K. L., Angstadt, M., Golden, J., Onyewuenyi, I., Popovska, A., & de Wit, H. (2008). Cannabinoid modulation of amygdala reactivity to social signals of threat in humans. *The Journal of Neuroscience, 28*(10), 2313-2319. https://doi.org/10.1523/JNEUROSCI.5603-07.2008

Phatale, V., Vaiphei, K. K., Jha, S., Patil, D., Agrawal, M., & Alexander, A. (2022). Overcoming skin barriers through advanced transdermal drug delivery approaches. *Journal of Controlled Release, 351*, 361-380. https://doi.org/10.1016/j.jconrel.2022.09.025

Pickard, H., Serge, H. A., & Foddy, B. (2015). Alternative models of addiction. *Frontiers in Psychiatry, 6* https://doi.org/10.3389/fpsyt.2015.00020

Piomelli, D., & Russo, E. B. (2016). The *Cannabis sativa* Versus *Cannabis indica* Debate: An Interview with Ethan Russo, MD. *Cannabis and cannabinoid research, 1*(1), 44–46. https://doi.org/10.1089/can.2015.29003.ebr

Pisani, A., Fezza, F., Galati, S., Battista, N., Napolitano, S., Finazzi-Agrò, A., ... & Maccarrone, M. (2005). High endogenous cannabinoid levels in the cerebrospinal fluid of untreated Parkinson's disease patients. *Annals of Neurology: Official Journal of the American Neurological Association and the Child Neurology Society, 57*(5), 777-779. https://doi.org/10.1002/ana.20462

Pisanti, S., & Bifulco, M. (2019). Medical cannabis: A plurimillennial history of an evergreen. *Journal of Cellular Physiology, 234*(6), 8342.

Pi-Sunyer, F. X., Aronne, L. J., Heshmati, H. M., Devin, J., Rosenstock, J., & RIO-North America Study Group, F. T. (2006). Effect of rimonabant, a cannabinoid-1 receptor blocker, on weight and cardiometabolic risk factors in overweight or obese patients: RIO-North America: a randomized controlled trial. *Jama, 295*(7), 761-775. https://doi.org/10.1001/jama.295.7.761

Pitman, R. K., Rasmusson, A. M., Koenen, K. C., Shin, L. M., Orr, S. P., Gilbertson, M. W., Milad, M. R., & Liberzon, I. (2012). Biological studies of post-traumatic stress disorder. *Nature Reviews Neuroscience, 13*(11), 769. https://doi.org/10.1038/nrn3339

Porter, A. C., Sauer, J., Knierman, M. D., Becker, G. W., Berna, M. J., Bao, J., Nomikos, G. G., Carter, P., Bymaster, F. P., Leese, A. B., & Felder, C. C. (2002). Characterization of a novel endocannabinoid, virodhamine, with antagonist activity at the CB1 receptor. *The Journal of Pharmacology and Experimental Therapeutics, 301*(3), 1020-1024. https://doi.org/10.1124/jpet.301.3.1020

Powles, T., Poele, R. t., Shamash, J., Chaplin, T., Propper, D., Joel, S., Oliver, T., & Liu, W. M. (2005). Cannabis-induced cytotoxicity in leukemic cell lines: The role of the cannabinoid receptors and the MAPK pathway. *Blood, 105*(3), 1214-1221. https://doi.org/10.1182/blood-2004-03-1182

Prince, M. A., & Conner, B. T. (2019). Examining links between cannabis potency and mental and physical health outcomes. *Behaviour Research and Therapy, 115*, 111-120. https://doi.org/10.1016/j.brat.2018.11.008

Preedy, V. R. (2020). Oxidative stress and dietary antioxidants. *Pathology*. Academic Press.

Prud'homme, M., & Cata, R. (2015). Cannabidiol as an intervention for addictive behaviors: A systematic review of the evidence. *Substance Abuse: Research and Treatment, 33*. https://doi.org/10.4137/SART.S25081.

Purcell, J. M., Passley, T. M., & Leheste, J. R. (2022). The cannabidiol and marijuana research expansion act: Promotion of scientific knowledge to prevent a national health crisis. *Lancet regional health. Americas, 14*, 100325. https://doi.org/10.1016/j.lana.2022.100325

Putterman, L. (2008). Agriculture, diffusion and development: Ripple effects of the neolithic revolution. *Economica, 75*(300), 729. https://doi.org/10.1111/j.1468-0335.2007.00652.x

Qian, Y., Gilliland, T. K., & Markowitz, J. S. (2020). The influence of carboxylesterase 1 polymorphism and cannabidiol on the hepatic metabolism of heroin. *Chemico-Biological Interactions, 316*https://doi.org/10.1016/j.cbi.2019.108914

Reid, M. (2020). A qualitative review of cannabis stigmas at the twilight of prohibition. *Journal of Cannabis Research, 2*(1)https://doi.org/10.1186/s42238-020-00056-8

Ren, G., Zhang, X., Li, Y., Ridout, K., Serrano-Serrano, M., Yang, Y., Liu, A., Ravikanth, G., Nawaz, M. A., Mumtaz, A. S., Salamin, N.,

& Fumagalli, L. (2021). Large-scale whole-genome resequencing unravels the domestication history of cannabis sativa. *Science Advances, 7*(29)https://doi.org/10.1126/sciadv.abg2286

Report to Congress, (2012). Mandatory minimum penalties in the federal criminal justice system. *Federal Sentencing Reporter, 24*(3), 185-192.

Reuter, P., & Pardo, B. (2017). Can new psychoactive substances be regulated effectively? an assessment of the british psychoactive substances bill. *Addiction, 112*(1), 25-31. https://doi.org/10.1111/add.13439

Reynolds, J. R. (1890). On the therapeutical uses and toxic effects of cannabis indica. *The Lancet, 135*(3473), 637-638. https://doi.org/10.1016/S0140-6736(02)18723-X

Richards, C. A., Jacobs, R. F., & Kaye, K. S. (2013). CDC calls for immediate action to control spread of CRE in hospitals. *Infectious Diseases in Children, 26*(4), 6-7

Richardson, D., Pearson, R. G., Kurian, N., Latif, M. L., Garle, M. J., Barrett, D. A., Kendall, D. A., Scammell, B. E., Reeve, A. J., & Chapman, V. (2008). Characterisation of the cannabinoid receptor system in synovial tissue and fluid in patients with osteoarthritis and rheumatoid arthritis. *Arthritis Research & Therapy, 10*, R43. https://doi.org/10.1186/ar2401

Richardson M. (2003). Understanding the structure and function of the skin. *Nursing times, 99*(31), 46–48. Macmillan Publishing Ltd.

Richey, J. M., Woolcott, O. O., Stefanovski, D., Harrison, L. N., Zheng, D., Lottati, M., Hsu, I. R., Kim, S. P., Kabir, M., Catalano, K. J., Chiu, J. D., Ionut, V., Kolka, C., Mooradian, V., & Bergman, R. N. (2009). Rimonabant prevents additional accumulation of visceral and subcutaneous fat during high-fat feeding in dogs. *The American Journal of Physiology, 296*(6), E1311 https://doi.org/10.1152/ajpendo.90972.2008

Richey, J. M., & Woolcott, O. (2017). Re-visiting the endocannabinoid system and its therapeutic potential in obesity and associated

diseases. *Current Diabetes Reports, 17*(10), 1-7.
https://doi.org/10.1007/s11892-017-0924-x

Rieder, S. A., Chauhan, A., Singh, U., Nagarkatti, M., & Nagarkatti, P. (2010). Cannabinoid-induced apoptosis in immune cells as a pathway to immunosuppression. *Immunobiology, 215*(8), 598–605. https://doi.org/10.1016/j.imbio.2009.04.001

Rindos, D. (1984). *The origins of agriculture: An evolutionary perspective.* Academic Press.

Robbins, I. P. (2018). Guns N' ganja: How federalism criminalizes the lawful use of marijuana. *U.C. Davis Law Review, 51*(5), 1783-1826.

Roberts, S. O., & Rizzo, M. T. (2021). The psychology of american racism. *American Psychologist, 76*(3), 475-487.
https://doi.org/10.1037/amp0000642

Robinson, T. E., & Berridge, K. C. (1993). The neural basis of drug craving: An incentive-sensitization theory of addiction. *Brain Research. Brain Research Reviews, 18*(3), 247-291.
https://doi.org/10.1016/0165-0173(93)90013-p

Robinson, P. D., Cools, P. D., Carlisi, B. A., Sahakian, P. D., & Drevets, (2012). Ventral striatum response during reward and punishment reversal learning in unmedicated major depressive disorder. *American Journal of Psychiatry: Official Journal of the American Psychiatric Association, 169*(2), 152-159.
https://doi.org/10.1176/appi.ajp.2011.11010137

Rock, E. M., Limebeer, C. L., Pertwee, R. G., Mechoulam, R., & Parker, L. A. (2021). Therapeutic potential of cannabidiol, cannabidiolic acid, and cannabidiolic acid methyl ester as treatments for nausea and vomiting. *Cannabis and Cannabinoid Research, 6*(4), 266-274.
https://doi.org/10.1089/can.2021.0041

Rohr, A. C., Wilkins, C. K., Clausen, P. A., Hammer, M., Nielsen, G. D., Wolkoff, P., & Spengler, J. D. (2002). Upper airway and pulmonary effects of oxidation products of (+)-alpha-pinene, d-limonene, and isoprene in BALB/c mice. *Inhalation toxicology, 14*(7), 663–684. https://doi.org/10.1080/08958370290084575

Romero-Sanchiz, P., Nogueira-Arjona, R., Pastor, A., Araos, P., Serrano, A., Boronat, A., Garcia-Marchena, N., Mayoral, F., Bordallo, A., Alen, F., Suárez, J., de la Torre, R., Pavón, F. J., & Rodríguez de Fonseca, F. (2019). Plasma concentrations of oleoylethanolamide in a primary care sample of depressed patients are increased in those treated with selective serotonin reuptake inhibitor-type antidepressants. *Neuropharmacology, 149*, 212-220. https://doi.org/10.1016/j.neuropharm.2019.02.026

Rosenblum, A., Marsch, L. A., Joseph, H., & Portenoy, R. K. (2008). Opioids and the treatment of chronic pain: Controversies, current status, and future directions. *Experimental and Clinical Psychopharmacology, 16*(5), 405-416. https://doi.org/10.1037/a0013628

Rosenkrantz, H., & Fleischman, R. W. (1978). Effects of cannabis on lungs. *Advances in the Biosciences, 22-23*, 279-299. https://doi.org/10.1016/b978-0-08-023759-6.50026-3

Roth-Deri, I., Friedman, A., Abraham, L., Lax, E., Flaumenhaft, Y., Dikshtein, Y., & Yadid, G. (2009). Antidepressant treatment facilitates dopamine release and drug seeking behavior in a genetic animal model of depression. *European Journal of Neuroscience, 30*(3), 485. https://doi.org/10.1111/j.1460-9568.2009.06840.x

Rush, B. (1814). *The drunkard's emblem, or, an inquiry into the effects of ardent spirits upon the human body and mind: With an account of the means of preventing, and of the remedies for curing them.* Harvard University Press.

Rush, B. (1816). *Extracts from dr. benjamin rush's inquiry into the effects of ardent spirits upon the human body and mind.* National Library of Medicine, Washington.

Russell, K., Cahill, M., & Duderstadt, K. G. (2019). Medical marijuana guidelines for practice: Health policy implications. *Journal of Pediatric*

Health Care, *33*(6), 722-726. https://doi.org/10.1016/j.pedhc.2019.07.010

Russo, E. (2001). Hemp for headache: An in-depth historical and scientific review of cannabis in migraine treatment. *Journal of Cannabis Therapeutics, 1*(2), 21-92

Russo E. B. (2008). Clinical endocannabinoid deficiency (CECD): Can this concept explain therapeutic benefits of cannabis in migraine, fibromyalgia, irritable bowel syndrome and other treatment-resistant conditions? *Neuro endocrinology letters, 25*(1-2), 31–39.

Russo, E. B. (2007). History of cannabis and its preparations in saga, science, and sobriquet. *Chemistry & Biodiversity, 4*(8), 1614-1648. https://doi.org/10.1002/cbdv.200790144

Russo, E. B. (2011). Taming THC: Potential cannabis synergy and phytocannabinoid-terpenoid entourage effects. *British Journal of Pharmacology, 163*(7), 1344-1364. https://doi.org/10.1111/j.1476-5381.2011.01238.x

Russo, E. B. (2018). Cannabis therapeutics and the future of neurology. *Frontiers in Integrative Neuroscience, 12* https://doi.org/10.3389/fnint.2018.00051

Russo E. B. (2019). The Case for the Entourage Effect and Conventional Breeding of Clinical Cannabis: No "Strain," No Gain. *Frontiers in plant science, 9*, 1969. https://doi.org/10.3389/fpls.2018.01969

Russo, E. B., Cuttler, C., Cooper, Z. D., Stueber, A., Whiteley, V. L., & Sexton, M. (2022). Survey of patients employing cannabigerol-predominant cannabis preparations: Perceived medical effects, adverse events, and withdrawal symptoms. *Cannabis and Cannabinoid Research, 7*(5), 706-716. https://doi.org/10.1089/can.2021.0058

Russo, E. B., Guy, G. W., & Robson, P. J. (2007). Cannabis, pain, and sleep: Lessons from therapeutic clinical trials of sativex, a cannabis-based medicine. *Chemistry & Biodiversity, 4*(8), 1729-1743. https://doi.org/10.1002/cbdv.200790150

Russo, E. B., Jiang, H. E., Li, X., Sutton, A., Carboni, A., del Bianco, F., Mandolino, G., Potter, D. J., Zhao, Y. X., Bera, S., Zhang, Y. B., Lü, E. G., Ferguson, D. K., Hueber, F., Zhao, L. C., Liu, C. J., Wang, Y. F., & Li, C. S. (2008). Phytochemical and genetic analyses of ancient cannabis from Central Asia. *Journal of experimental botany, 59*(15), 4171–4182. https://doi.org/10.1093/jxb/ern260

Russo, E. B., & Marcu, J. (2017). Cannabis pharmacology: The usual suspects and a few promising leads. *Advances in Pharmacology (San Diego, Calif.), 80*, 67-134. https://doi.org/10.1016/bs.apha.2017.03.004

Ryan, D., Drysdale, A. J., Lafourcade, C., Pertwee, R. G., & Platt, B. (2009). Cannabidiol targets mitochondria to regulate intracellular Ca2+ levels. *The Journal of Neuroscience, 29*(7), 2053-2063. https://doi.org/10.1523/JNEUROSCI.4212-08.2009

Sales, A. J., Crestani, C. C., Guimarães, F. S., & Joca, S. R. L. (2018). Antidepressant-like effect induced by Cannabidiol is dependent on brain serotonin levels. *Progress in neuro-psychopharmacology & biological psychiatry, 86*, 255–261. https://doi.org/10.1016/j.pnpbp.2018.06.002

Sallaberry, C. A., & Astern, L. (2018). The endocannabinoid system, our universal regulator. *Journal of Young Investigators, 34*(6), 48–55. https://doi.org/10.22186/jyi.34.5.48-55

Sallan, S. E., Cronin, C., Zelen, M., & Zinberg, N. E. (1980). Antiemetics in patients receiving chemotherapy for cancer - A randomized comparison of delta-9-tetrahydrocannabinol and prochlorperazine. *New England Journal of Medicine, 302*(3), 135-138. https://doi.org/10.1056/NEJM198001173020302

Salter, P. S., Adams, G., & Perez, M. J. (2018). Racism in the structure of everyday worlds: A cultural-psychological perspective. *Current Directions in Psychological Science, 27*(3), 150-155. https://doi.org/10.1177/0963721417724239

Salzman, C., Kochansky, G., Van Der Kolk, B., & Shader, R. (1977). The effect of marijuana on small group process. *The American Journal* of *Drug and Alcohol Abuse, 4*(2), 251-255. https://doi.org/10.3109/00952997709002763

Salzman, C., Van der Kolk, B. A., & Shader, R. I. (1976). Marijuana and hostility in a small-group setting. *The American Journal of Psychiatry, 133*(9), 1029-1033. https://doi.org/10.1176/ajp.133.9.1029

Santini, A., Cammarata, S. M., Capone, G., Ianaro, A., Tenore, G. C., Pani, L., & Novellino, E. (2018). Nutraceuticals: Opening the debate for a regulatory framework. *British Journal of Clinical Pharmacology, 84*(4), 659-672. https://doi.org/10.1111/bcp.13496

Santos, J. M. O., Costa, A. C., Dias, T. R., Satari, S., Costa E Silva, M. P., da Costa, R.,M.Gil, & Medeiros, R. (2021). Towards drug repurposing in cancer cachexia: Potential targets and candidates. *Pharmaceuticals (Basel, Switzerland), 14*(11) https://doi.org/10.3390/ph14111084

Sareen J. (2014). Posttraumatic stress disorder in adults: impact, comorbidity, risk factors, and treatment. *Canadian journal of psychiatry. Revue canadienne de psychiatrie*, *59*(9), 460–467. https://doi.org/10.1177/070674371405900902

Sarnes, E., Crofford, L., Watson, M., Dennis, G., Kan, H., & Bass, D. (2011). Incidence and US costs of corticosteroid-associated adverse events: A systematic literature review. *Clinical Therapeutics, 33*(10), 1413-1432. https://doi.org/10.1016/j.clinthera.2011.09.009

Sarris, J., Sinclair, J., Karamacoska, D., Davidson, M., & Firth, J. (2020). Medicinal cannabis for psychiatric disorders: A clinically-focused systematic review. *BMC Psychiatry, 20*(1), 1-14. https://doi.org/10.1186/s12888-019-2409-8

Sartim, A. G., Guimarães, F. S., & Joca, S. R. L. (2016). Antidepressant-like effect of cannabidiol injection into the ventral medial prefrontal cortex—Possible involvement of 5-HT1A and CB1 receptors. *Behavioural Brain Research, 303*, 218-227. https://doi.org/10.1016/j.bbr.2016.01.033

Sasaki, K., & Yoshizaki, F. (2015). Investigation into hippocampal nerve cell damage through the mineralocorticoid receptor in mice. *Molecular Medicine Reports,,* 7211. https://doi.org/10.3892/mmr.2015.4406

Schiepers, O. J. G., Wichers, M. C., & Maes, M. (2005). Cytokines and major depression. *Progress in Neuropsychopharmacology & Biological Psychiatry, 29*(2), 201

Schmidt, M, S. (2017). US Drug Enforcement Administration head resigns, saying Donald Trump disrespects law. *The Sydney Morning Herald.* Available from: https://www.smh.com.au/world/us-drug-enforcement-administration-head resigns-saying-donald-trump-disrespects-law-20170927-gyphav.html

Schneider, J. W. (1978). Deviant drinking as disease: Alcoholism as a social accomplishment. *Social Problems, 25*(4), 361-372. https://doi.org/10.1525/sp.1978.25.4.03a00020

Schnelle, M., Grotenhermen, F., Reif, M., & Gorter, R. W. (1999). [Results of a standardized survey on the medical use of cannabis products in the german-speaking area]. *Forschende Komplementarmedizin, 6 Suppl 3*, 28-36. https://doi.org/10.1159/000057154

Schou, T. M., Joca, S., Wegener, G., & Bay-Richter, C. (2021). Psychiatric and neuropsychiatric sequelae of COVID-19—A systematic review. *Brain, Behavior, and Immunity, 97*, 328-348. https://doi.org/10.1016/j.bbi.2021.07.018

Schwarz, A. M., Keresztes, A., Bui, T., Hecksel, R. J., Peña, A., Lent, B., ... & Streicher, J. M. (2023). Terpenes from Cannabis sativa Induce Antinociception in Mouse Chronic Neuropathic Pain via Activation of Spinal Cord Adenosine A2A Receptors. *bioRxiv.*

Schwarz, H., Blanco, F. J., & Lotz, M. (1994). Anadamide, an endogenous cannabinoid receptor agonist inhibits lymphocyte proliferation and induces apoptosis. *Journal of Neuroimmunology, 55*(1), 107-115. https://doi.org/10.1016/0165-5728(94)90152-x

Scott, C., Neira Agonh, D., & Lehmann, C. (2022). Antibacterial effects of phytocannabinoids. *Life (2075-1729), 12*(9), 1394–N.PAG. https://doi.org/10.3390/life12091394

Scutt, A., & Williamson, E. M. (2007). Cannabinoids stimulate fibroblastic colony formation by bone marrow cells indirectly via CB2 receptors. *Calcified Tissue International, 80*(1), 50-59. https://doi.org/10.1007/s00223-006-0171-7

Sea, Y. L., Gee, Y. J., Lal, S. K., & Choo, W. S. (2023). Cannabis as antivirals. *Journal of Applied Microbiology, 134*(1)https://doi.org/10.1093/jambio/lxac036

Seillier, A., Martinez, A. A., & Giuffrida, A. (2013). Phencyclidine-induced social withdrawal results from deficient stimulation of cannabinoid CB_1 receptors: Implications for schizophrenia. *Neuropsychopharmacology: Official Publication of the American College of Neuropsychopharmacology, 38*(9), 1816-1824. https://doi.org/10.1038/npp.2013.81

Serpell, M., Ratcliffe, S., Hovorka, J., Schofield, M., Taylor, L., Lauder, H., & Ehler, E. (2014). A double-blind, randomized, placebo-controlled, parallel group study of THC/CBD spray in peripheral neuropathic pain treatment. *European Journal of Pain (London, England), 18*(7), 999-1012. https://doi.org/10.1002/j.1532-2149.2013.00445.x

Servadio, M., Melancia, F., Manduca, A., di Masi, A., Schiavi, S., Cartocci, V., Pallottini, V., Campolongo, P., Ascenzi, P., & Trezza, V. (2016). Targeting anandamide metabolism rescues core and associated autistic-like symptoms in rats prenatally exposed to valproic acid. *Translational Psychiatry, 6*(9), e902. https://doi.org/10.1038/tp.2016.182

Sexton, M., Cuttler, C., Finnell, J. S., & Mischley, L. K. (2016). A cross-sectional survey of medical cannabis users: Patterns of use and perceived efficacy. *Cannabis and Cannabinoid Research, 1*(1), 131-138. https://doi.org/10.1089/can.2016.0007

Shahbazi, F., Grandi, V., Banerjee, A., & Trant, J. F. (2020). Cannabinoids and cannabinoid receptors: The story so far. *iScience, 23*(7) https://doi.org/10.1016/j.isci.2020.101301

Shanks, K. G., Clark, W., & Behonick, G. (2016). Death associated with the use of the synthetic cannabinoid ADB-FUBINACA. *Journal of Analytical Toxicology, 40*(3), 236–239. https://doi.org/10.1093/jat/bkv142

Shannon, S., & Opila-Lehman, J. (2015). Cannabidiol Oil for Decreasing Addictive Use of Marijuana: A Case Report. *Integrative medicine (Encinitas, Calif.), 14*(6), 31–35.

Sharkey, K. A., & Wiley, J. W. (2016). The role of the endocannabinoid system in the brain-gut axis. *Gastroenterology, 151*(2), 252-266. https://doi.org/10.1053/j.gastro.2016.04.015

Sharma, C., Al Kaabi, J. M., Nurulain, S. M., Goyal, S. N., Kamal, M. A., & Ojha, S. (2016). Polypharmacological properties and therapeutic potential of β-caryophyllene: A dietary phytocannabinoid of pharmaceutical promise. *Current Pharmaceutical Design, 22*(21), 3237-3264. https://doi.org/10.2174/1381612822666160311115226

Shen, M., & Thayer, S. A. (1999). Δ9-Tetrahydrocannabinol acts as a partial agonist to modulate glutamatergic synaptic transmission between rat hippocampal neurons in culture. *Molecular Pharmacology, 55*(1), 8-13. https://doi.org/10.1124/mol.55.1.8

Shen, B., Zhang, D., Zeng, X., Guan, L., Yang, G., Liu, L., Huang, J., Li, Y., Hong, S., & Li, L. (2022). Cannabidiol inhibits methamphetamine-induced dopamine release via modulation of the DRD1-MeCP2-BDNF-TrkB signaling pathway. *Psychopharmacology, 239*(5), 1521-1537. https://doi.org/10.1007/s00213-021-06051-y

Shereen Lai, S. M., Sook, Y. L., Nelson Jeng, Y. C., Bey, H. G., Wen-Nee Tan, & Kooi, Y. K. (2022). Plant terpenoids as the promising source of cholinesterase inhibitors for anti-AD therapy. *Biology, 11*(2), 307. https://doi.org/10.3390/biology11020307

Shier, A. C., Reichenbacher, T., Ghuman, H. S., & Ghuman, J. K. (2013). Pharmacological treatment of attention deficit hyperactivity disorder in children and adolescents: Clinical strategies. *Journal of Central Nervous System Disease,* (5), 1-17. https://doi.org/10.4137/JCNSD.S6691

Shinjyo, N., & Di Marzo, V. (2013). The effect of cannabichromene on adult neural stem/progenitor cells. *Neurochemistry International, 63*(5), 432-437. https://doi.org/10.1016/j.neuint.2013.08.002

Skaper, S. D., & Di Marzo, V. (2012). Endocannabinoids in nervous system health and disease: The big picture in a nutshell. *Philosophical Transactions of the Royal Society of London.Series B, Biological Sciences, 367*(1607), 3193-3200. https://doi.org/10.1098/rstb.2012.0313

Shonesy, B. C., Winder, D. G., Patel, S., & Colbran, R. J. (2015). The initiation of synaptic 2-AG mobilization requires both an increased supply of diacylglycerol precursor and increased postsynaptic calcium. *Neuropharmacology, 91 Suppl. C*, 57-62. https://doi.org/10.1016/j.neuropharm.2014.11.026

Shugerman, J. H. (2014). The creation of the department of justice: Professionalization without civil rights or civil service. *Stanford Law Review, 66*(1), 121-172.

Siegel, J. D., Rhinehart, E., Jackson, M., & Chiarello, L. (2007). Management of multidrug-resistant organisms in health care settings, 2006. *AJIC: American Journal of Infection Control, 35*(10), S165-S193. https://doi.org/10.1016/j.ajic.2007.10.006

Sieniawska, E., Swatko-Ossor, M., Sawicki, R., Skalicka-Woźniak, K., & Ginalska, G. (2017). Natural terpenes influence the activity of antibiotics against isolated mycobacterium tuberculosis. *Medical Principles and Practice: International Journal of the Kuwait University, Health Science Centre, 26*(2), 108-112. https://doi.org/10.1159/000454680

Siklos-Whillans, J., Bacchus, A., & Manwell, L. A. (2021). A scoping review of the use of cannabis and its extracts as potential harm reduction strategies: Insights from preclinical and clinical research. *International Journal of Mental Health and Addiction, 19*(5), 1527-1550. https://doi.org/10.1007/s11469-020-00244-w

Silva, E. A. d., Junior, Medeiros, W. M. B., Torro, N., Sousa, J. M. M. d., Almeida, I. B. C. M. d., Costa, F. B. d., Pontes, K. M., Nunes,

E. L. G., Rosa, M. D. d., & Albuquerque, K. L. G. D. d. (2022). Cannabis and cannabinoid use in autism spectrum disorder: A systematic review. *Trends in Psychiatry and Psychotherapy, 44*, e20200149. https://doi.org/10.47626/2237-6089-2020-0149

Silva-Correa, C., Campos-Reyna, J., Villarreal-La Torre, V. E., Calderón-Peña, A. A., Sagástegui-Guarniz, W. A., Guerrero-Espino, L., González-Siccha, A. D., Aspajo-Villalaz, C., González-Blas, M. V., Cruzado-Razco, J., & Hilario-Vargas, J. (2021). Potential neuroprotective activity of essential oils in memory and learning impairment. *Pharmacognosy Journal, 13*(5), 1312-1322. https://doi.org/10.5530/pj.2021.13.166

Silva, E. A. D., Junior, Medeiros, W. M. B., Torro, N., Sousa, J. M. M., Almeida, I. B. C. M., Costa, F. B. D., Pontes, K. M., Nunes, E. L. G., Rosa, M. D. D., & Albuquerque, K. L. G. D. (2022). Cannabis and cannabinoid use in autism spectrum disorder: a systematic review. *Trends in psychiatry and psychotherapy, 44*, e20200149. https://doi.org/10.47626/2237-6089-2020-0149

Singh, D., Dilnawaz, F., & Sahoo, S. K. (2020). Challenges of moving theranostic nanomedicine into the clinic. *Nanomedicine (London, England), 15*(2), 111-114. https://doi.org/10.2217/nnm-2019-0401

Sink, K. M., Holden, K. F., & Yaffe, K. (2005). Pharmacological treatment of neuropsychiatric symptoms of dementia: A review of the evidence. *JAMA, the Journal of the American Medical Association, 293*(5), 596.

Slaughter, J. B. (1988). *Marijuana prohibition in the united states: History and analysis of a failed policy*. Columbia University School of Law.

Śledziński, P., Zeyland, J., Słomski, R., & Nowak, A. (2018). The current state and future perspectives of cannabinoids in cancer biology. *Cancer Medicine, 7*(3), 765-775. https://doi.org/10.1002/cam4.1312

Smaga, I., Bystrowska, B., Gawliński, D., Przegaliński, E., & Filip, M. (2014). The endocannabinoid/endovanilloid system and depression. *Current Neuropharmacology, 12*(5), 462-474. https://doi.org/10.2174/1570159X12666140923205412

Smalheiser, N. R. (2019). A neglected link between the psychoactive effects of dietary ingredients and consciousness-altering drugs. *Frontiers in Psychiatry, 10* https://doi.org/10.3389/fpsyt.2019.00591

Small, E. (2015). Evolution and classification of cannabis sativa (marijuana, hemp) in relation to human utilization. *The Botanical Review, 81*(3), 189-294. https://doi.org/10.1007/s12229-015-9157-3

Smith, J. A. (1996). Beyond the divide between cognition and discourse: Using interpretative phenomenological analysis in health psychology. *Psychology & Health, 11*(2), 261-271. https://doi.org/10.1080/08870449608400256

Solomon, R. (2020). Racism and its effect on cannabis research. *Cannabis & Cannabinoid Research, 5*(1), 2-5. https://doi.org/10.1089/can.2019.0063

Somvanshi, R. K., Zou, S., Kadhim, S., Padania, S., Hsu, E., & Kumar, U. (2022). Cannabinol modulates neuroprotection and intraocular pressure: A potential multi-target therapeutic intervention for glaucoma. *BBA - Molecular Basis of Disease, 1868*(3) https://doi.org/10.1016/j.bbadis.2021.166325

Sorenson, J. L. & Johannesse, C. L. (2009). *World trade and biological exchanges before 1492.* Bloomington, Ind.: iUniverse.

Spindle, T. R., Zamarripa, C. A., Russo, E., Pollak, L., Bigelow, G., Ward, A. M., Tompson, B., Sempio, C., Shokati, T., Klawitter, J., Christians, U., & Vandrey, R. (2024). Vaporized D-limonene selectively mitigates the acute anxiogenic effects of Δ9-tetrahydrocannabinol in healthy adults who intermittently use cannabis. *Drug and Alcohol Dependence, 257* https://doi.org/10.1016/j.drugalcdep.2024.111267

Spindle, T. R., Zamarripa, C. A., Russo, E., Pollak, L., Bigelow, G., Ward, A. M., Tompson, B., Sempio, C., Shokati, T., Klawitter, J., Christians, U., & Vandrey, R. (2024). Vaporized D-limonene selectively mitigates the acute anxiogenic effects of Δ9-tetrahydrocannabinol in healthy adults who intermittently use cannabis. *Drug and Alcohol Dependence, 257*https://doi.org/10.1016/j.drugalcdep.2024.111267

Sreepian, A., Popruk, S., Nutalai, D., Phutthanu, C., & Sreepian,

P. M. (2022). Antibacterial activities and synergistic interaction of citrus essential oils and limonene with gentamicin against clinically isolated methicillin-resistant staphylococcus aureus. *TheScientificWorldJournal, 2022*, 8418287. https://doi.org/10.1155/2022/8418287

Stachnik, J., & Gabay, M. (2010). Emerging role of aripiprazole for treatment of irritability associated with autistic disorder in children and adolescents. *Adolescent Health, Medicine and Therapeutics, 1*, 105-114. https://doi.org/10.2147/AHMT.S9819

Stampanoni Bassi, M., Sancesario, A., Morace, R., Centonze, D., & Iezzi, E. (2017). Cannabinoids in parkinson's disease. *Cannabis and Cannabinoid Research, 2*(1), 21-29. https://doi.org/10.1089/can.2017.0002

Stanley, C. P., Wheal, A. J., Randall, M. D., & O'Sullivan, S. E. (2013). Cannabinoids alter endothelial function in the zucker rat model of type 2 diabetes. *European Journal of Pharmacology, 720*(1-3), 376. https://doi.org/10.1016/j.ejphar.2013.10.002

Stasiłowicz, A., Tomala, A., Podolak, I., & Cielecka-Piontek, J. (2021). Cannabis sativa L. as a natural drug meeting the criteria of a multitarget approach to treatment. *International Journal of Molecular Sciences, 22*(2)https://doi.org/10.3390/ijms22020778

Stein, D. (2017). Pew analysis finds no relationship between drug imprisonment and drug problems. *Pew Trusts*. Available at https://www.pewtrusts.org/en/research-and-analysis/speeches-and-testimony/2017/06/pew-analysis-finds-no-relationship-between-drug-imprisonment-and-drug-problems

Stein, L. (1964). Self-stimulation of the brain and the central stimulant action of amphetamine. *Federation Proceedings, 23*, 836-850.

Steinberg, M., Shao, H., Zandi, P., Lyketsos, C. G., Welsh-Bohmer, K., Norton, M. C., Breitner, J. C. S., Steffens, D. C., Tschanz, J. T., & Investigators, C. C. (2008). Point and 5-year period prevalence of neuropsychiatric symptoms in dementia: The cache county study. *International Journal of Geriatric Psychiatry, 23*(2), 170-177. https://doi.org/10.1002/gps.1858

Stith, S. S., Diviant, J. P., Brockelman, F., Keeling, K., Hall, B., Lucern, S., & Vigil, J. M. (2020). Alleviative effects of cannabis flower on migraine and headache. *Journal of Integrative Medicine, 18*(5), 416-424. https://doi.org/10.1016/j.joim.2020.07.004

Stone, N. L., Murphy, A. J., England, T. J., & O'Sullivan, S.,E. (2020). A systematic review of minor phytocannabinoids with promising neuroprotective potential. *British Journal of Pharmacology, 177*(19), 4330-4352. https://doi.org/10.1111/bph.15185

Stone, M., & Robert, J. (2020). The cannabis catch-22: DEA suffocates cannabis research because we don't understand cannabis. *Southern University Law Review, 47*(2), 383-422

Strand, N. H., Maloney, J., Kraus, M., Wie, C., Turkiewicz, M., Gomez, D. A., Adeleye, O., & Harbell, M. W. (2023). Cannabis for the treatment of fibromyalgia: A systematic review. *Biomedicines, 11*(6) https://doi.org/10.3390/biomedicines11061621

Stringer, R. J., & Maggard, S. R. (2016). Reefer madness to marijuana legalization: Media exposure and american attitudes toward marijuana (1975-2012). *Journal of Drug Issues, 46*(4), 428. https://doi.org/10.1177/0022042616659762

Su, T., Yan, Y., Li, Q., Ye, J., & Pei, L. (2021). Endocannabinoid system unlocks the puzzle of autism treatment via microglia. *Frontiers in Psychiatry, 12* https://doi.org/10.3389/fpsyt.2021.734837

Substance Abuse and Mental Health Services Administration. (2019). Key substance use and mental health indicators in the United States: Results from the 2018 National Survey on Drug Use and Health. *HHS Publication No. PEP19-5068, NSDUH Series H-54, 170*, 51-58.

Sun, M., Mainland, B. J., Ornstein, T. J., Mallya, S., Fiocco, A. J., Sin, G. L., Shulman, K. I., & Herrmann, N. (2018). The association between cognitive fluctuations and activities of daily living and quality of life among institutionalized patients with dementia. *International Journal of Geriatric Psychiatry, 33*(2), e280-e285. https://doi.org/10.1002/gps.4788

Sun, Y., Qu, Y., & Zhu, J. (2021). The relationship between inflammation and post-traumatic stress disorder. *Frontiers in Psychiatry, 12*https://doi.org/10.3389/fpsyt.2021.707543

Surendran, S., Qassadi, F., Surendran, G., Lilley, D., & Heinrich, M. (2021). Myrcene-what are the potential health benefits of this flavouring and aroma agent? *Frontiers in Nutrition, 8*, 699666. https://doi.org/10.3389/fnut.2021.699666

Suryadevara, U., Bruijnzeel, D. M., Nuthi, M., Jagnarine, D. A., Tandon, R., & Bruijnzeel, A. W. (2017). Pros and cons of medical cannabis use by people with chronic brain disorders. *Current Neuropharmacology, 15*(6), 800-814. https://doi.org/10.2174/1570159X14666161101095325

Szkudlarek, H. J., Rodríguez-Ruiz, M., Hudson, R., De Felice, M., Jung, T., Rushlow, W. J., & Laviolette, S. R. (2021). THC and CBD produce divergent effects on perception and panic behaviours via distinct cortical molecular pathways. *Progress in Neuro-Psychopharmacology and Biological Psychiatry, 104*, 110029. https://doi.org/10.1016/j.pnpbp.2020.110029

Tait, R. J., Caldicott, D., Mountain, D., Hill, S. L., & Lenton, S. (2016). A systematic review of adverse events arising from the use of synthetic cannabinoids and their associated treatment. *Clinical toxicology (Philadelphia, Pa.), 54*(1), 1–13. https://doi.org/10.3109/15563650.2015.1110590

Tanda, G., Pontieri, F. E., & Di Chiara, G. (1997). Cannabinoid and heroin activation of mesolimbic dopamine transmission by a common mu1 opioid receptor mechanism. *Science, 276* (5321), 2048. https://doi.org/10.1126/science.276.5321.2048

Tanimura, A., Yamazaki, M., Hashimotodani, Y., Uchigashima, M., Kawata, S., Abe, M., Kita, Y., Hashimoto, K., Shimizu, T., Watanabe, M., Sakimura, K., & Kano, M. (2010). The endocannabinoid 2-arachidonoylglycerol produced by diacylglycerol lipase α mediates retrograde suppression of synaptic transmission. *Neuron, 65*(3), 320-327. https://doi.org/10.1016/j.neuron.2010.01.021

Tashkin, D. P., Shapiro, B. J., Lee, Y. E., & Harper, C. E. (1975). Effects of smoked marijuana in experimentally induced asthma. *The American Review of Respiratory Disease, 112*(3), 377-386. https://doi.org/10.1164/arrd.1975.112.3.377

Trautman, L. J., Seaborn, P., Sulkowski, A., Mayer, D., & Luttrell, R. T. I.,II. (2021). Cannabis at the crossroads: A transdisciplinary analysis and policy prescription. *Oklahoma City University Law Review, 45*, 125-188.

Thiele, E. A., Marsh, E. D., French, J. A., Mazurkiewicz-Beldzinska, M., Benbadis, S. R., Joshi, C., Lyons, P. D., Taylor, A., Roberts, C., & Sommerville, K. (2018). Cannabidiol in patients with seizures associated with lennox-gastaut syndrome (GWPCARE4): A randomised, double-blind, placebo-controlled phase 3 trial. *Lancet, 391*(10125), 1085-1096. https://doi.org/10.1016/S0140-6736(18)30136-3

Thomas, A., Baillie, G. L., Phillips, A. M., Razdan, R. K., Ross, R. A., & Pertwee, R. G. (2007). Cannabidiol displays unexpectedly high potency as an antagonist of CB1 and CB2 receptor agonists in vitro. *British Journal of Pharmacology, 150*(5), 613-623. https://doi.org/10.1038/sj.bjp.0707133

Thomas, K. H., Martin, R. M., Potokar, J., Pirmohamed, M., & Gunnell, D. (2014). Reporting of drug induced depression and fatal and non-fatal suicidal behaviour in the UK from 1998 to 2011. *BMC Pharmacology & Toxicology, 15*, 54. https://doi.org/10.1186/2050-6511-15-54

Timler, A., Bulsara, C., Bulsara, M., Vickery, A., Smith, J., & Codde, J. (2020). Use of cannabinoid-based medicine among older residential care recipients diagnosed with dementia: study protocol for a double-blind randomised crossover trial. *Trials, 21*, 1-11. https://doi.org/10.1186/s13063-020-4085-x

Toft, H., Lien, L., Neupane, S. P., Abebe, D. S., Tilden, T., Wampold, B. E., & Bramness, J. G. (2020). Cytokine concentrations are related to level of mental distress in inpatients not using anti-inflammatory drugs. *Acta neuropsychiatrica, 32*(1), 23-31.

Tomida, I., Azuara-Blanco, A., House, H., Flint, M., Pertwee, R. G., & Robson, P. J. (2006). Effect of sublingual application of cannabinoids on intraocular pressure: A pilot study. *Journal of Glaucoma, 15*(5), 349-353. https://doi.org/10.1097/01.ijg.0000212260.04488.60

Ton, J. M. N. C., Gerhardt, G. A., Friedemann, M., Etgen, A. M., Rose, G. M., Sharpless, N. S., & Gardner, E. L. (1988). The effects of !D–9-tetrahydrocannabinol on potassium-evoked release of dopamine in the rat caudate nucleus: An in vivo electrochemical and in vivo microdialysis study. *Brain Research, 451*(1-2), 59-68. https://doi.org/10.1016/0006-8993(88)90749-4

Torjesen, I. (2013). Antimicrobial resistance presents an "apocalyptic" threat similar to that of climate change, CMO warns. *Bmj, 346*, f1597. https://doi.org/10.1136/bmj.f1597

Torkamaneh, D., & Jones, A. M. P. (2022). Cannabis, the multibillion-dollar plant that no genebank wanted. *Genome, 65*(1), 1. https://doi.org/10.1139/gen-2021-0016

Trezza, V., Campolongo, P., Manduca, A., Morena, M., Palmery, M., Vanderschuren, L., & Cuomo, V. (2012). Altering endocannabinoid neurotransmission at critical developmental ages: Impact on rodent emotionality and cognitive performance. *Frontiers in Behavioral Neuroscience, 6*https://doi.org/10.3389/fnbeh.2012.00002

Trezza, V., Damsteegt, R., Manduca, A., Petrosino, S., Van Kerkhof, L. W. M., Pasterkamp, R. J., Zhou, Y., Campolongo, P., Cuomo, V., Di Marzo, V., & Vanderschuren, L. J. M. J. (2012). Endocannabinoids in amygdala and nucleus accumbens mediate social play reward in adolescent rats. *The Journal of Neuroscience, 32*(43), 14899-14908. https://doi.org/10.1523/JNEUROSCI.0114-12.2012

Trillou, C. R., Arnone, M., Delgorge, C., Gonalons, N., Keane, P., Maffrand, J., & Soubrié, P. (2003). Anti-obesity effect of SR141716, a CB1 receptor antagonist, in diet-induced obese mice. *American*

Journal of Physiology: Regulatory, Integrative & Comparative Physiology, 53(2), R345. https://doi.org/10.1152/ajpregu.00545.2002

Trillou, C. R., Delgorge, C., Menet, C., Arnone, M., & Soubrié, P. (2004). CB1 cannabinoid receptor knockout in mice leads to leanness, resistance to diet-induced obesity and enhanced leptin sensitivity. *International Journal of Obesity & Related Metabolic Disorders, 28*(4), 640-648. https://doi.org/10.1038/sj.ijo.0802583

Troutt, W. D., & DiDonato, M. D. (2017). Carbonyl compounds produced by vaporizing cannabis oil thinning agents. *Journal of Alternative & Complementary Medicine, 23*(11), 879-884. https://doi.org/10.1089/acm.2016.0337

Tudge, L., Williams, C., Cowen, P. J., & McCabe, C. (2014). Neural effects of cannabinoid CB1 neutral antagonist tetrahydrocannabivarin on food reward and aversion in healthy volunteers. *The International Journal of neuropsychopharmacology, 18*(6) https://doi.org/10.1093/ijnp/pyu094

Tye, K. M., Mirzabekov, J. J., Warden, M. R., Ferenczi, E. A., Tsai, H. C., Finkelstein, J., Kim, S. Y., Adhikari, A., Thompson, K. R., Andalman, A. S., Gunaydin, L. A., Witten, I. B., & Deisseroth, K. (2013). Dopamine neurons modulate neural encoding and expression of depression-related behaviour. *Nature, 493*(7433), 537–541. https://doi.org/10.1038/nature11740

Tysnes, O. B., & Storstein, A. (2017). Epidemiology of parkinson's disease. *Journal of Neural Transmission (Vienna, Austria : 1996), 124*(8), 901-905. https://doi.org/10.1007/s00702-017-1686-y

Uguen, M., Perrin, D., Belliard, S., Ligneau, X., Beardsley, P. M., Lecomte, J. M., & Schwartz, J. C. (2013). Preclinical evaluation of the abuse potential of Pitolisant, a histamine H₃ receptor inverse agonist/antagonist compared with Modafinil. *British journal of pharmacology, 169*(3), 632–644. https://doi.org/10.1111/bph.12149

Underhill, A. P. (1997). Current issues in chinese neolithic archaeology. *Journal of World Prehistory, 11*(2), 103-160. https://doi.org/10.1007/bf02221203

Ungerleider, J. T., Andrysiak, T., Fairbanks, L., Goodnight, J., Sarna, G., & Jamison, K. (1982). Cannabis and cancer chemotherapy: A comparison of oral delta-9-THC and prochlorperazine. *Cancer, 50*(4), 636-645. https://doi.org/10.1002/1097-0142(19820815)50:4

United States Department of Health and Human Services (2003). Patent # 6630507 Cannabinoids as antioxidants and neuroprotectants. Washington, DC, USA.

Urban, N. B. L., Kegeles, L. S., Slifstein, M., Xu, X., Martinez, D., Sakr, E., Castillo, F., Moadel, T., O'Malley, S. S., Krystal, J. H., & Abi-Dargham, A. (2010). Sex differences in striatal dopamine release in young adults after oral alcohol challenge: A positron emission tomography imaging study with [^{11}C]raclopride. *Biological Psychiatry, 68*(8), 689-696. https://doi.org/10.1016/j.biopsych.2010.06.005

USDA, (1943). *Hemp for Victory!* [Video] Hemp for Victory (full version) - Bing video

Valeri, A., Chiricosta, L., D'Angiolini, S., Pollastro, F., Salamone, S., & Mazzon, E. (2023). Cannabichromene induces neuronal differentiation in NSC-34 cells: Insights from transcriptomic analysis. *Life (2075-1729), 13*(3), 742. https://doi.org/10.3390/life13030742

VanDolah, H. J., Bauer, B. A., & Mauck, K. F. (2019). Clinicians' guide to cannabidiol and hemp oils. *Mayo Clinic Proceedings, 94*(9), 1840. https://doi.org/10.1016/j.mayocp.2019.01.003

Varmus, H. (2018). Humanism for Innovation. *Innovative Research in Life Sciences: Pathways to Scientific Impact, Public Health Improvement, and Economic Progress*, 133.

Verhoeckx, K. C., Korthout, H. A., van Meeteren-Kreikamp, A. P., Ehlert, K. A., Wang, M., van der Greef, J., ... & Witkamp, R. F. (2006). Unheated Cannabis sativa extracts and its major compound THC-acid have potential immuno-modulating properties not mediated by CB1 and CB2 receptor coupled pathways. *International immunopharmacology, 6*(4), 656-665. https://doi.org/10.1016/j.intimp.2005.10.002 Vermeij, E. A., Koenders, M. I., Blom, A. B.,

Arntz, O. J., Bennink, M. B., van den Berg, W. B., van Lent, P. L. E. M., & van de Loo, F. A. J. (2014). In vivo molecular imaging of cathepsin and matrix metalloproteinase activity discriminates between arthritic and osteoarthritic processes in mice. *Molecular Imaging, 13*(2), 1-10. https://doi.org/10.2310/7290.2014.00001

Vespermann, K. A. C., Paulino, B. N., Barcelos, M. C. S., Pessôa, M. G., Pastore, G. M., & Molina, G. (2017). Biotransformation of α- and β-pinene into flavor compounds. *Applied Microbiology and Biotechnology, 101*(5), 1805-1817. https://doi.org/10.1007/s00253-016-8066-7

Viana, M. d. B., Aquino, P. E. A. d., Estadella, D., Ribeiro, D. A., & Viana, G. S. d. B. (2022). Cannabis sativa and cannabidiol: A therapeutic strategy for the treatment of neurodegenerative diseases? *Medical Cannabis and Cannabinoids, 5*(1), 207. https://doi.org/10.1159/000527335

Vieira, A. J., Beserra, F. P., Souza, M. C., Totti, B. M., & Rozza, A. L. (2018). Limonene: Aroma of innovation in health and disease. *Chemico-Biological Interactions, 283*, 97-106. https://doi.org/10.1016/j.cbi.2018.02.007

Viveros, M. P., Marco, E. M., & File, S. E. (2005). Endocannabinoid system and stress and anxiety responses. *Pharmacology, Biochemistry and Behavior, 81*(2), 331-342. https://doi.org/10.1016/j.pbb.2005.01.029

Volkow, N. D., Fowler, J. S., & Wang, G. J. (2003). The addicted human brain: insights from imaging studies. *The Journal of clinical investigation, 111*(10), 1444–1451. https://doi.org/10.1172/JCI18533

Volkow, N. D., Wang, G. J., Fowler, J. S., Logan, J., Schlyer, D., Hitzemann, R., Lieberman, J., Angrist, B., Pappas, N., MacGregor, R., & et. al. (1994). Imaging endogenous dopamine competition with [11C]raclopride in the human brain. *Synapse (New York, N.Y.), 16*(4), 255-262. https://doi.org/10.1002/syn.890160402

Volkow, N. D., Wang, G. J., Fowler, J. S., Logan, J., Gatley, S. J., Gifford, A., Hitzemann, R., Ding, Y. S., & Pappas, N. (1999).

Prediction of reinforcing responses to psychostimulants in humans by brain dopamine D2 receptor levels. *The American journal of psychiatry, 156*(9), 1440–1443. https://doi.org/10.1176/ajp.156.9.1440

Vora, L. K., Gholap, A. D., Hatvate, N. T., Naren, P., Khan, S., Chavda, V. P., Balar, P. C., Gandhi, J., & Khatri, D. K. (2024). Essential oils for clinical aromatherapy: A comprehensive review. *Journal of Ethnopharmacology, 330* https://doi.org/10.1016/j.jep.2024.118180

Vuic, B., Milos, T., Tudor, L., Konjevod, M., Nikolac Perkovic, M., Jazvinscak Jembrek, M., Nedic Erjavec, G., & Svob Strac, D. (2022). Cannabinoid CB2 receptors in neurodegenerative proteinopathies: New insights and therapeutic potential. *Biomedicines, 10*(12) https://doi.org/10.3390/biomedicines10123000

Wadden, T. A. (1993). Treatment of obesity by moderate and severe caloric restriction. results of clinical research trials. *Annals of Internal Medicine, 119*(7), 688-693. https://doi.org/10.7326/0003-4819-119-7_part_2-199310011-00012

Wagner, H., & Ulrich-Merzenich, G. (2009). Synergy research: Approaching a new generation of phytopharmaceuticals. *Phytomedicine, 16*(2), 97-110. https://doi.org/10.1016/j.phymed.2008.12.018

Walitt, B., Klose, P., Fitzcharles, M., Phillips, T., & Häuser, W. (2016). Cannabinoids for fibromyalgia. *The Cochrane Database of Systematic Reviews, 7*, CD011694. https://doi.org/10.1002/14651858.CD011694.pub2

Wallace, M., Schulteis, G., Atkinson, J. H., Wolfson, T., Lazzaretto, D., Bentley, H., Gouaux, B., & Abramson, I. (2007). Dose-dependent effects of smoked cannabis on capsaicin-induced pain and hyperalgesia in healthy volunteers. *Anesthesiology, 107*(5), 785-796. https://doi.org/10.1097/01.anes.0000286986.92475.b7

Wallace, M. S., Marcotte, T. D., Umlauf, A., Gouaux, B., & Atkinson, J. H. (2015). Efficacy of inhaled cannabis on painful diabetic neuropathy. *Journal of Pain, 16*(7), 616-627. https://doi.org/10.1016/j.jpain.2015.03.008

Waller, D. G., & Sampson, A. (2017). *Medical pharmacology and therapeutics : Medical pharmacology and therapeutics E-book*. Elsevier.

Wang, C., Lu, H., Gu, W., Wu, N., Zhang, J., Zuo, X., Li, F., Wang, D., Dong, Y., Wang, S., Liu, Y., Bao, Y., & Hu, Y. (2019). The development of yangshao agriculture and its interaction with social dynamics in the middle yellow river region, china. *Holocene, 29*(1), 173-180. https://doi.org/10.1177/0959683618804640

Waters, S. E. (2019). Punishing the immoral other: Penal substitutionary logic in the war on drugs. *Pastoral Psychology, 68*(5), 533-548. https://doi.org/10.1007/s11089-018-0836-y

Watkins, L. E., Sprang, K. R., & Rothbaum, B. O. (2018). Treating PTSD: A review of evidence-based psychotherapy interventions. *Frontiers in Behavioral Neuroscience, 12*https://doi.org/10.3389/fnbeh.2018.00258

Wehling, M. (2014). Non-steroidal anti-inflammatory drug use in chronic pain conditions with special emphasis on the elderly and patients with relevant comorbidities: Management and mitigation of risks and adverse effects. *European Journal of Clinical Pharmacology, 70*(10), 1159-1172. https://doi.org/10.1007/s00228-014-1734-6

Wei, D., Dinh, D., Lee, D., Li, D., Anguren, A., Moreno-Sanz, G., Gall, C. M., & Piomelli, D. (2016). Enhancement of anandamide-mediated endocannabinoid signaling corrects autism-related social impairment. *Cannabis and Cannabinoid Research*, (1) https://doi.org/10.1089/can.2015.0008

Wei, D., Lee, D., Cox, C. D., Karsten, C. A., Peñagarikano, O., Geschwind, D. H., Gall, C. M., & Piomelli, D. (2015). Endocannabinoid signaling mediates oxytocin-driven social reward. *Proceedings of the National Academy of Sciences of the United States of America, 112*(45), 14084-14089. https://doi.org/10.1073/pnas.1509795112

Wenzel, J. M., & Joseph, F. C. (2014). Endocannabinoid-dependent modulation of phasic dopamine signaling encodes external and internal reward-predictive cues. *Frontiers in Psychiatry, 5*https://doi.org/10.3389/fpsyt.2014.00118

Wenzel, R. P. (2004). Nosocomial bloodstream infections in US hospitals: Analysis of 24,179 cases from a prospective nationwide surveillance study. *Clinical Infectious Diseases, 39*(3), 309-317.

Weston-Green, K., Clunas, H., & Jimenez Naranjo, C. (2021). A review of the potential use of pinene and linalool as terpene-based medicines for brain health: Discovering novel therapeutics in the flavours and fragrances of cannabis. *Frontiers in Psychiatry, 12* https://doi.org/10.3389/fpsyt.2021.583211

Whalley, B. (2007). Cannabis and epilepsy: From recreational abuse to therapeutic use. *University of Reading.* www.societyofbiology.org/images/ben-whalley.pdf.

Wilker, S., Pfeiffer, A., Elbert, T., Ovuga, E., Karabatsiakis, A., Krumbholz, A., Thieme, D., Schelling, G., & Kolassa, I. (2016). Endocannabinoid concentrations in hair are associated with PTSD symptom severity. *Psychoneuroendocrinology, 67*, 198-206. https://doi.org/10.1016/j.psyneuen.2016.02.010

Williams, D. R., & Williams-Morris, R. (2000). Racism and mental health: The african american experience. *Ethnicity & Health, 5*(3), 243-268. https://doi.org/10.1080/713667453

Williams, S. J., Hartley, J. P., & Graham, J. D. (1976). Bronchodilator effect of delta1-tetrahydrocannabinol administered by aerosol of asthmatic patients. *Thorax, 31*(6), 720-723. https://doi.org/10.1136/thx.31.6.720

Wimo, A., Seeher, K., Cataldi, R., Cyhlarova, E., Dielemann, J. L., Frisell, O., Guerchet, M., Jönsson, L., Malaha, A. K., Nichols, E., Pedroza, P., Prince, M., Knapp, M., & Dua, T. (2023). The worldwide costs of dementia in 2019. *Alzheimer's & Dementia : The Journal of the Alzheimer's Association, 19*(7), 2865-2873. https://doi.org/10.1002/alz.12901

Wink, L. K., Plawecki, M. H., Erickson, C. A., Stigler, K. A., & McDougle, C. J. (2010). Emerging drugs for the treatment of symptoms associated with autism spectrum disorders. *Expert Opinion on Emerging Drugs, 15*(3), 481-494. https://doi.org/10.1517/14728214.2010.487860

Wilsey, B., Marcotte, T., Deutsch, R., Gouaux, B., Sakai, S., & Donaghe, H. (2013). Low-dose vaporized cannabis significantly improves neuropathic pain. *Journal of Pain, 14*(2), 136-148. https://doi.org/10.1016/j.jpain.2012.10.009

Wise, R. A., & Bozarth, M. A. (1987). A psychomotor stimulant theory of addiction. *Psychological Review, 94*(4), 469–492. https://doi.org/10.1037/0033-295X.94.4.469

Wolf, S. A., Bick-Sander, A., Fabel, K., Leal-Galicia, P., Tauber, S., Ramirez-Rodriguez, G., Müller, A., Melnik, A., Waltinger, T. P., Ullrich, O., & Kempermann, G. (2010). Cannabinoid receptor CB1 mediates baseline and activity-induced survival of new neurons in adult hippocampal neurogenesis. *Cell Communication & Signaling, 8*, 12-25. https://doi.org/10.1186/1478-811X-8-12

Wroński, A., Jarocka-Karpowicz, I., Stasiewicz, A., & Skrzydlewska, E. (2023). Phytocannabinoids in the pharmacotherapy of psoriasis. *Molecules, 28*(3), 1192. https://doi.org/10.3390/molecules28031192

Wu, J., Peng, W., Yi, J., Wu, Y., Chen, T., Wong, K., & Wu, J. (2014). Chemical composition, antimicrobial activity against staphylococcus aureus and a pro-apoptotic effect in SGC-7901 of the essential oil from toona sinensis (A. juss.) roem. leaves. *Journal of Ethnopharmacology, 154*(1), 198-205. https://doi.org/10.1016/j.jep.2014.04.002

Xu, L., Li, X., Zhang, Y., Ding, M., Sun, B., Su, G., & Zhao, Y. (2021). The effects of linalool acupoint application therapy on sleep regulation. *RSC Advances, 11*(11), 5896-5902. https://doi.org/10.1039/d0ra09751a

Yang, H., Zhou, J., & Lehmann, C. (2016). GPR55 - a putative "type 3" cannabinoid receptor in inflammation. *Journal of Basic and Clinical Physiology and Pharmacology, 27*(3), 297-302. https://doi.org/10.1515/jbcpp-2015-0080

Yasa Ozturk, G., & Bashan, I. (2021). The effect of aromatherapy with lavender oil on the health-related quality of life in patients with

fibromyalgia. *Journal of Food Quality, 2021*
https://doi.org/10.1155/2021/9938630

Yewale, C., Tandel, H., Patel, A., & Misra, A. (2021). Chapter 5 - polymers in transdermal drug delivery. In A. Misra, & A. Shahiwala (Eds.), *Applications of polymers in drug delivery (second edition)* (pp. 131-158). Elsevier.
https://doi.org/10.1016/B978-0-12-819659-5.00005-7

Yüksek, A., Havuz, S. G., Karaduman, N. Ş., Şimsek, H., & Honca, M. (2021). An investigation on the antimicrobial activity of hemp fiber and fabrics againist common nasocomial infection agents. *Journal of Cukurova Anesthesia and Surgical Sciences, 5*(2), 137-144. https://doi.org/10.36516/jocass.1125626

Zajicek, J., Fox, P., Sanders, H., Wright, D., Vickery, J., Nunn, A., Thompson, A., Zajicek, J., Fox, P., Sanders, H., Wright, D., Vickery, J., Nunn, A., & Thompson, A. (2003). Cannabinoids for treatment of spasticity and other symptoms related to multiple sclerosis (CAMS study): Multicentre randomised placebo-controlled trial. *Lancet, 362*(9395), 1517-1526.
https://doi.org/10.1016/s0140-6736(03)14738-1

Zamora-Mendoza, L., Guamba, E., Miño, K., Romero, M. P., Levoyer, A., Alvarez-Barreto, J., Machado, A., & Alexis, F. (2022). Antimicrobial properties of plant fibers. *Molecules, 27*(22), 7999.
https://doi.org/10.3390/molecules27227999

Zhang, L., Yang, Z., Fan, G., Ren, J., Yin, K., & Pan, S. (2019). Antidepressant-like effect of citrus sinensis (L.) osbeck essential oil and its main component limonene on mice. *Journal of Agricultural and Food Chemistry, 67*(50), 13817-13828.
https://doi.org/10.1021/acs.jafc.9b00650

Zhang, Y., Li, J., Zhao, Y., Wu, X., Li, H., Yao, L., Zhu, H., & Zhou, H. (2017). Genetic diversity of two neolithic populations provides evidence of farming expansions in north china. *Journal of Human Genetics, 62*(2), 199-204. https://doi.org/10.1038/jhg.2016.107

Zhang, Z., Guo, S., Liu, X., & Gao, X. (2015). Synergistic antitumor effect of α-pinene and β-pinene with paclitaxel against non-small-cell lung carcinoma (NSCLC). *Drug Research, 65*(4), 214-218. https://doi.org/10.1055/s-0034-1377025

Zinderman, C. E., Conner, B., Malakooti, M. A., LaMar, J. E., Armstrong, A., & Bohnkert, B. K. (2004). Community-acquired methicillin-resistant staphylococcus aureus among military recruits. *Emerging Infectious Diseases, 10*(5), 941-944. https://doi.org/10.3201/eid1005.030604

Zoerner, A. A., Gutzki, F., Batkai, S., May, M., Rakers, C., Engeli, S., Jordan, J., & Tsikas, D. (2011). Quantification of endocannabinoids in biological systems by chromatography and mass spectrometry: A comprehensive review from an analytical and biological perspective. *Biochimica Et Biophysica Acta, 1811*(11), 706-723. https://doi.org/10.1016/j.bbalip.2011.08.004

Zou, S., & Kumar, U. (2018). Cannabinoid receptors and the endocannabinoid system: Signaling and function in the central nervous system. *International Journal of Molecular Sciences, 19*(3), 833. https://doi.org/10.3390/ijms19030833

Zurier, R. B., & Burstein, S. H. (2016). Cannabinoids, inflammation, and fibrosis. *FASEB Journal: Official Publication of the Federation of American Societies for Experimental Biology, 30*(11), 3682-3689. https://doi.org/10.1096/fj.201600646R

www.ingramcontent.com/pod-product-compliance
Lightning Source LLC
Chambersburg PA
CBHW020454030426
42337CB00011B/110